PAPUA NEW GUINEA

Science

Grade 8

Student Book

Kenneth Rouse

OXFORD

Level 8, 737 Bourke Street, Docklands, Victoria 3008, Australia.

Oxford University Press is a department of the University of Oxford. It furthers the University's objective of excellence in research, scholarship, and education by publishing worldwide in

Oxford New York

Auckland Cape Town Dar es Salaam Hong Kong Karachi
Kuala Lumpur Madrid Melbourne Mexico City Nairobi
New Delhi Shanghai Taipei Toronto

With offices in

Argentina Austria Brazil Chile Czech Republic France Greece
Guatemala Hungary Italy Japan Poland Portugal Singapore
South Korea Switzerland Thailand Turkey Ukraine Vietnam

OXFORD is a trademark of Oxford University Press in the UK and in certain other countries

First published 2008
Reprinted 2008, 2009, 2010, 2013 (twice), 2014 (twice), 2015, 2016, 2017, 2018, 2020, 2021, 2023, 2026

ISBN 978 0 19 555518 9

Typeset by Pier Vido
Illustrated by Uramina & Nelson Ltd
Printed in China by Golden Cup Printing Co. Ltd
Oxford University Press Australia & New Zealand is committed to sourcing paper responsibly.

Acknowledgments
The author and publisher wish to thank the following copyright holders for granting permission to reproduce their material.
Corbis/Jay Dickman p. 16; Corbis /Wayne Lawler p. 61; Crawford House Publishing p.17 (right); CSIRO p.47; Fotolia pp. 71, 110 (top right), 126, 114, 127, 134; Istockphoto pp 57, 92, 109, 113 (bottom); Jupiter Images pp. 77 (bottom right), 130; 111; NASA pp. 124, 125; Photolibrary pp. 41, 73 (top right), 75, 77, 110 (left & bottom right), 130; Ken Rouse, pp. 58, 108; Smithsonian Institution Historical Images Collection, courtesy General Electric, p. 91.
Every effort has been made to trace the original source of copyright material contained in this book. The publisher would be pleased to hear from copyright holders to rectify any errors or omissions.

Contents

Working scientifically

Chapter summary

In this chapter you will have an opportunity to:

- find out more about ways of doing investigations that are fair and how to use controls in your investigations
- collect and record information as accurately as possible for each investigation that you carry out
- make conclusions based on the information that you have collected and what you have learned from your experience
- decide if your answers are fair to the questions asked
- find out how to collect information and make good decisions about the way that you can use science in everyday life.

Syllabus references

Strand: Working scientifically

Sub-strand: Working scientifically

Outcomes: **8.1.1** Identify the role of science in the global environment and apply scientific methods to create solutions to problems.

Key facts

- People want to understand the world around them and find out the answers to many different questions.
- A good way to find out more about why things happen and how they work is to carry out scientific investigations, which usually means that we try to find the answer to a question.
- When we carry out an investigation we must make sure that we are asking a question that will help us to understand and then choose a way to find an answer to that question.
- When we carry out investigations we must also try to make sure that we are being fair, and this usually means that we must use controls.
- Each investigation will be different because it has a different purpose, but we must always be accurate in the way we collect and record information.
- The conclusions that we make must be based on the information that we have collected and be linked to the purpose of the investigation.
- When we think that we have found an answer we must look at the question again and decide if the answer is sensible or reasonable.
- When we use the scientific method to collect reliable information and answer questions we can make good decisions about the way that we can use science in everyday life.

Being observant (PD) (SS)

Some people are more observant than others but we can all train ourselves to be good observers by being interested in what is happening and paying careful attention.

If you want to be good at investigating and solving problems then you need to be observant and try to understand what is happening by thinking about the things that you observe. This means that you must notice the things around you and pay attention to the details.

Smells should be tested carefully by fanning a small amount towards you with your hand.

All five senses are important for making observations—seeing, hearing, feeling, smelling and tasting. However, we sometimes need to be careful with some of our senses so that we do not hurt ourselves. For example:

- looking directly at a bright object like the Sun can damage our eyes
- being too close to loud music or a loud noise like an engine can damage our ears
- touching something very hot can burn our skin
- smelling or breathing in some **gases** can make us sick or even kill us
- tasting **solids** or **liquids** that are poisonous can make us sick or even kill us.

When you smell something it is best to use your hand to fan a small amount of the gas towards you. You should not taste something unless you know that it is safe to do so.

For you to try

1 Look at the drawing below. List the differences between the two men and what they are carrying. Discuss your results with other students. Does everyone agree on the differences?

2 Look at the drawing below and try to work out what it means.

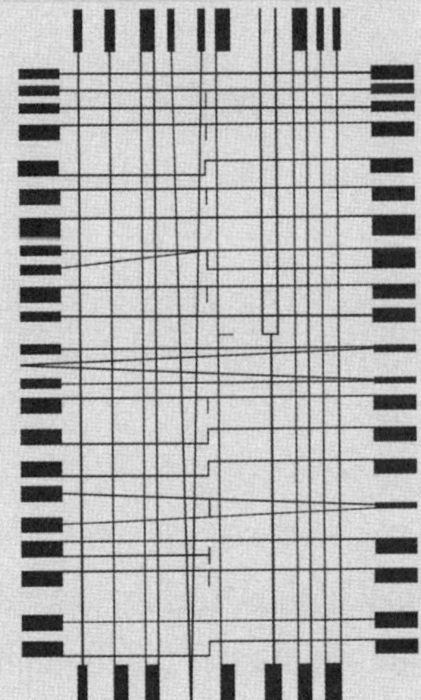

You will need to turn the page on its side to find the first part of the secret message. Then you can find the second part of the message when you turn the page the right way round.

3 Copy and complete the following table. For each item listed, put a tick in the column to show which sense or senses you would use to make observations of that item. A

Item	Sight	Hearing	Touch	Smell	Taste
bicycle					
mustard leaf					
rice					
aeroplane					
rock					
soup					
body spray					

4 Investigation: Which fruit is that?

a Collect a piece of cloth for a blindfold, a pair of gloves and a variety of fruits that are in season, such as banana, orange, lemon, lime, pomelo, breadfruit, custard apple, mango etc.

b Ask one student to wear the blindfold, then give them the fruit in any order and ask them to name the pieces of fruit. Record your results in a table. How many did they get correct?

c Repeat this with other students.

d Now blindfold another student and ask them to wear gloves so that they cannot feel the skin of the fruit. They cannot squeeze the fruit but they can hold it and smell it.

(continued over)

e Again, ask them to name the pieces of fruit. Record your results in a table. How many did they get correct? Different students can try this.

f Finally, repeat the experiment with a student wearing the blindfold and the gloves, but this time they are not allowed to smell the fruit. Again, record your results in a table. How many did they get correct?

g Write up a scientific report to explain what you did and answer the following questions:

- Does it help you to use more senses in this investigation?
- How important is it to use all your senses?
- Which students were the most observant?

5 Two students have had a water fight and the teacher has caught them. Drops of water have been splashed on the blackboard, on the desk and on the floor. The shape of the water drops is shown in the diagram below:

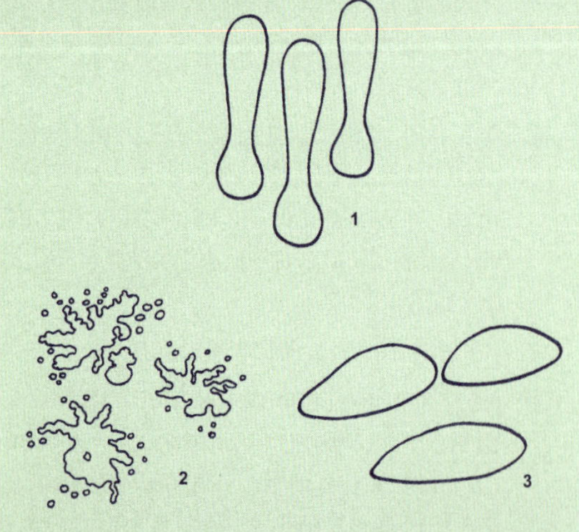

Using your observation skills which drop would be found:

a on the floor where it fell

b on the floor where it was thrown

c on the blackboard.

6 After a bullet has been fired from a gun we can find a mark on the back of the bullet. Because each gun is slightly different, this mark can be used to work out which gun fired the bullet. The diagram below shows the mark on the bullet from the scene of a crime at the top and the pattern on the bullets from six guns below. Which gun fired the bullet at the top?

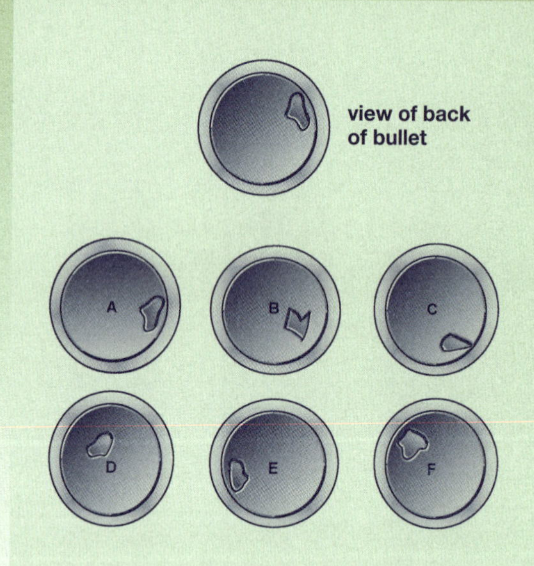

7 Work in small groups to discuss the following:

Imagine that you are trying to find the cause of pollution into a river or a bay. List the qualities you would need in order to be a good investigator. Which qualities would be most important? Share your list with other groups.

Doing experiments and using controls

A good way to find more about why things happen and how they work is to carry out scientific investigations, which usually means that we try to find the answer to a question. When we carry out an investigation we must make sure that we are asking a suitable question that will help us to understand and then choose a way to find an answer to that question.

When we carry out investigations or experiments we must also try to make sure that we are being **fair**, and this often means that we must use **controls**. There are usually two parts to the experiment which we do at the same time. One part is called the **test** or experiment, and the other is called the control. To be fair, we must treat the test or experiment and the control in exactly the same way but we make one change to the experiment. Then, when we compare the results of the test with the results from the control, we can say that any difference in the result is because of the different conditions that we applied to the test and the control. If we change more than one of the conditions of the experiment at the same time then we will not know which of the conditions caused the difference in the results. In other words, when we carry out a fair experiment we can interpret the difference in the results to help us make a conclusion. We must also be careful that the conclusion follows from the results.

So when we carry out an investigation we should follow a number of steps, and each one must lead to the next. For example:

1 ask a suitable question (what do you want to find out?)

2 choose suitable methods that are fair (how will you answer the question?)

3 get some results (what happened?)

4 make conclusions relating to the question (what do the results mean?)

Doing a fair experiment

The following example shows the way that we can carry out a proper and fair experiment using a test and a control.

Some students were trying to find out the best conditions for **seeds** to grow well. Both Group A and Group B had plenty of seeds and both were able to use any equipment that they needed. The reports of their experiments are shown below:

Group A

Aim To find out the best conditions for seeds to grow well.

Apparatus Seeds, plastic containers, water, shade house made from bush materials.

Method Many seeds were planted in each of three containers and given the following treatments.

Container	Treatment
Experiment one	Placed out in full sunlight and given water occasionally
Experiment two	Placed in the deep shade and given water occasionally
Control	Placed in part shade and given regular watering

(continued over)

Results

Container	Appearance of seedlings
Experiment one	Plants looked green and healthy but wilted
Experiment two	Plants looked pale and weak but did not wilt
Control	Plants grew well and did not wilt

Conclusion The seedlings from this plant grow best when they are in full sun.

Group B

Aim To find out the light conditions in which seedlings grow best.

Apparatus Seeds, plastic containers, water, shade house made from bush materials, ruler.

Method A few seeds were planted in fifteen containers. Five containers were then given each of the following treatments.

Containers	Treatment
Experiment one (five containers)	Placed out in full sunlight and given one can of water each day
Experiment two (five containers)	Placed in the deep shade and given one can of water each day
Control (five containers)	Placed in part shade and given one can of water each day

Results

Container	Appearance of seedlings	Height of seedlings
Experiment one	Green and healthy, no wilting	20 cm
Experiment two	Pale and weak, no wilting	10 cm
Control	Green and healthy, no wilting	30 cm

Conclusion The seeds from this plant seem to grow best when they are in part shade. Seedlings in deep shade become pale and week. Seedlings appear to be healthy when growing in the full sunlight but they are not as tall.

Was it a fair experiment?

Did both groups carry out a fair experiment?

Do their conclusions follow from their results?

Group A concluded that seedlings from this plant grew best in full sunlight because they were green and healthy. However, they gave the three pots different amounts of water and sunlight so they are not able to make any definite conclusions because they cannot be confident about the reasons for the results. Their conclusion does not follow from their method and result.

Also, they put many seeds in the same pot so there may have been competition between the seedlings. They did not measure the height of the seedlings but used only the appearance of the seedlings.

Group B chose a better title for their investigation, deciding to find out about the importance of sunlight. They put fewer seeds but used more containers so there was less likely to be competition between the seedlings and less of a problem if one or two seedlings died. They changed only one thing at a time, which was the amount of sunlight that the seedlings received. They did not change the amount of the water. They also measured the height of the seedlings as well as describing their appearance and so there were two different parts to their results including making measurements. It is then fair to say that

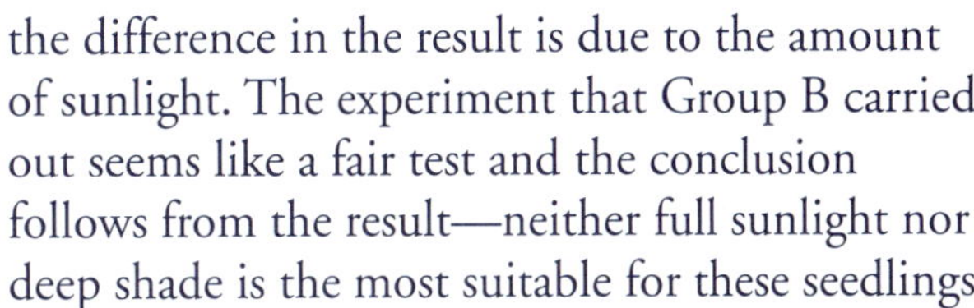

the difference in the result is due to the amount of sunlight. The experiment that Group B carried out seems like a fair test and the conclusion follows from the result—neither full sunlight nor deep shade is the most suitable for these seedlings.

For you to try

1. Write some notes for Group A to explain to them how their experiment could be improved. Remember to give reasons for your comments. **L**
2. **Investigation: To find the light conditions in which seedlings grow best** **SR**
   - a Collect some seeds, plastic containers, water, a shade house made from bush materials, and a ruler. Each group should try a different type of seed.
   - b Carry out an experiment similar to the one carried out by Group B above.
   - c Write a scientific report of your experiment to say what you did and what you found out.
   - d At the end of the experiment, compare your results with those of other groups and suggest reasons for any differences.

Confidence in our results

When the experiment is fair, somebody else should also be able to repeat the experiment and get the same results, or similar results. When different people repeat the same experiment and get similar results we can feel more confident about our experiments and the conclusions that we make. We can then feel sure that we understand the reasons why something is happening and that nothing is being missed.

When we carry out fair experiments the information that we obtain is more likely to be reliable. This means that the information is more likely to be honest and true so that we can believe it.

When we think that we have found an answer we must look at the question again and decide if the answer is sensible or reasonable. For example, imagine your friends from another school tell you that they have been carrying out some experiments to find the fastest runners in their school. They measured out 100 metres and had several heats and a final race. They tell you that in the final race the first three runners all ran the 100 metres in ten seconds.

Do you believe them? Does this result sound reasonable? How do you decide?

If you have no idea how long it takes to run 100 metres then you need to find out before you can comment on the results of your friends. For example, you might do the following:

- Try to run 100 metres yourself and measure the time taken or measure the time it takes

Timing a 100-metre race

your friends to run this distance. Is it close to ten seconds?

- Talk to someone who knows more about the time it takes to run 100 metres. What do they say?
- Look up the records to find out the fastest times in the world for men and women. For example, you might look up the results for the Olympic Games, the World Championships or the South Pacific Games.

If you did these things you would find that the fastest man in the world can run 100 metres in just under ten seconds and the fastest woman, in just over ten seconds. It would be very difficult for you and your friends to run 100 metres in ten seconds. The results are more likely to be between 3 and 17 seconds. So does this mean that your friends have got some of the fastest runners in the world at their school? This is very unlikely. Their result does not seem sensible or reasonable and so we cannot believe it. There must be some other explanation.

For you to try

1 Using the information in the example above about the time taken to run 100 metres, suggest some reasons for the results which your friends obtained. How could they check their results?

2 The table at the top of page 11 gives a summary of the way that the population of PNG has changed since Independence in 1975.

Population of PNG

Year	Population	Comments
1975 (Independence)	3 000 000	an estimate, not based on a census
2000	5 190 786	results from the census—number of people counted
2005	5 887 000	an estimate—the population seems to have doubled in about 30 years
?	?	
?	?	

M

Choose two dates in the future and make an estimate to say what you think the population will be. Remember that the number must be sensible or reasonable.

Showing data

When we carry out an investigation we collect information, which is also called **data**. The methods used in some experiments mean that we must take measurements and so we end up with a list of numbers. We then have to find a way of showing this information or these data in our results so that it is easy to see and understand. One of the most important things that we do is to look for patterns in the results.

Drawing tables and graphs

M

Tables and graphs are a way of sorting information into groups so that we can quickly see the information and can study it more easily.

The following example shows the way that we use tables and graphs to show information more easily. Notice the different ways that the information is displayed.

When she was on holiday Moana recorded in her diary some of the things that she was doing:

Moana's secret holiday diary

On Monday morning I listened to some music then I went to visit my friend's house before spending one and a half hours working in the garden. On Tuesday morning we played volleyball near the okari tree and then in the afternoon I worked in my mother's garden for an hour and then went to help my aunty in her garden for an hour. On Wednesday a group of us went swimming and fishing in the morning and after lunch I worked in the garden for about an hour. In the evening we went to the church for youth fellowship and on the way home one of the boys was teasing me. My friend says boys usually tease the girls that they like the best! On Thursday I had promised to help my mother, my aunty and my grandmother. I spent one and a half hours helping my mother clear her new garden, thirty minutes helping my aunt with her weeding and thirty minutes with my grandmother harvesting sweet potatoes. I was really tired after that! On Friday I helped my mother harvest enough food so that we do not need to go to the garden on the weekend. That took one hour to finish.

It is a little bit difficult to pick out all the different things that Moana did from her diary but one thing we can do is organise some of the information in a table as follows:

Time Moana spent on gardening

Day	Time (hours)
Monday	1.5
Tuesday	2
Wednesday	1
Thursday	2.5
Friday	1

Note the following rules about making a table:

- The table has a title or heading.
- The time spent is shown in hours. The minutes have been converted to fractions of an hour so that the times can be compared easily.
- The units (hours) are written at the top of the column. You don't have to write the units after each number.
- The table has neat lines drawn with a ruler and the numbers are in **vertical** lines.
- The table gives a summary of the information. It does not tell us which garden Moana worked in or what other things Moana did besides gardening. The table records only the time spent on gardening.
- If Moana did not work in the garden on one of these days then a zero would be recorded in the time column. If Moana forgot to record the time she spent gardening then a dash (–) would be written in the column. Do not leave blanks in a table.

Column graph

M

Another way to show the time that Moana spent working in the garden is to use a column graph or histogram. It is like a picture and allows us to see the information from the table more easily. A graph usually shows two sets of numbers or measurements that are related to each other. One set of numbers is shown on a line along the bottom and the other set is shown on a line on the side. The bottom line is also called the **horizontal axis** and the line on the side is called the **vertical** axis. Most graphs have two axes.

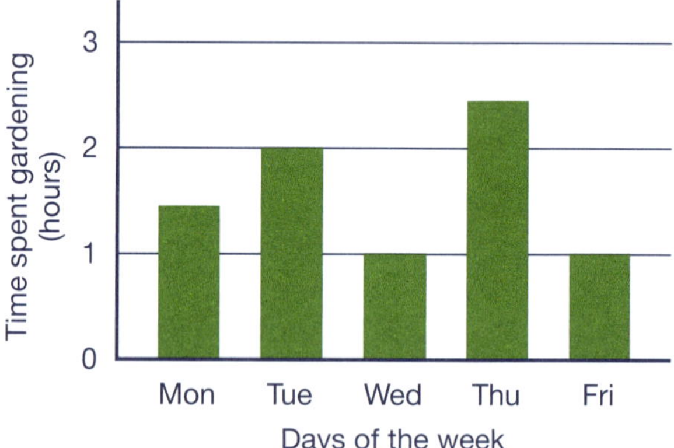

Note the following rules about making column graphs:

- The graph has a heading or title that explains the graph.
- Each axis has a title and units.
- The scale on each axis is even. There is the same distance between the days on the horizontal axis and between the number of hours on the vertical axis.
- The graph is drawn neatly using a ruler.
- The columns are the same thickness and the same distance apart.

For you to try

1 A class of 30 students was given a test which was out of 20 marks. The highest mark was 20 and the lowest was 11. The marks are shown below: **A**

20	14	19	14	18	16
19	18	20	18	13	18
15	17	14	16	19	14
18	16	19	11	18	17
17	19	15	18	16	18

Record the information in a table like the one below:

Test score	Number of students
20	
19	
18	
etc.	

Draw a column graph to show the data. The test score should be on the horizontal axis and the number of students should be on the vertical axis. Remember to give the graph a heading or title.

2 Look at the histogram below and answer the questions that follow. **A**

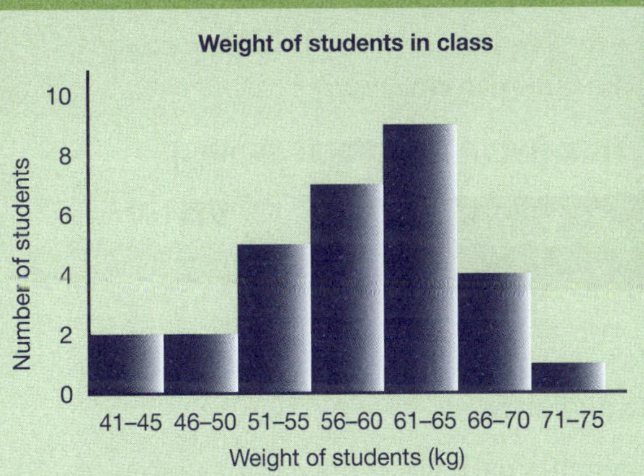

a How many students weigh between 56 and 60 kg?

b What is the most common weight of students in this class?

c What could be the smallest weight in this class?

d What could be the weight of the heaviest student in this class?

e How many students are in this class?

f List two pieces of information that this column graph does not tell you about the students.

More graphs

Two other ways of showing information in a way that is easy to see are bar graphs and line graphs.

Bar graphs

A bar graph is useful to show information that has fractions or percentages. For example, we can show the different amounts of the gases that are found in air.

Amount of gases in air

Gas	Percentage (%)
Nitrogen	78
Oxygen	21
Other gases	1

A bar graph shows these amounts as different columns in a bar.

Notes to help you draw bar graphs:

- Choose a suitable scale. For example, if you make your bar graph 10 cm long, that makes 1 mm equal to 1% and it is easier to show different percentages.

A bar graph

Line graphs

A column graph shows the data in a series of columns. A line graph has a line joining the points where the middle of the top of the columns would be. These points are called data points. Instead of drawing the columns you draw the data points and a line joining the points. Most of the graphs that we draw in **science** are line graphs.

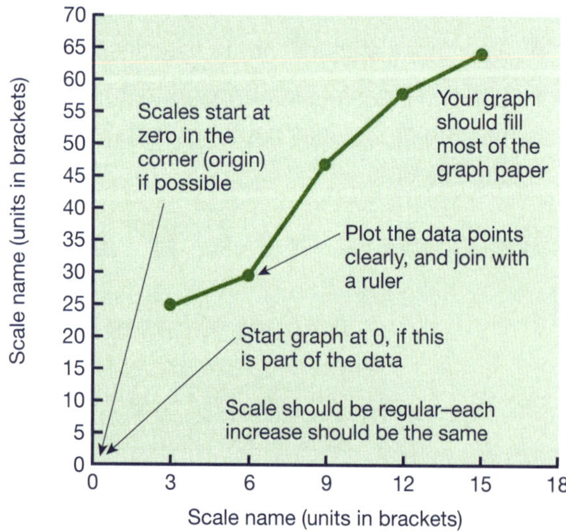

How to draw a line graph

- Remember to include a key to show what the different parts of the graph stand for.

Notes to help you draw line graphs:

- Remember to include a suitable title that explains the graph.
- Choose a suitable scale with regular intervals.
- You do not need to have both axes start at 0—you must look at the numbers first before you decide, especially the lowest value.
- Your graph should fill most of the graph paper.
- Label each axis clearly.
- Show the units on each axis.
- Plot the data points clearly.
- Draw the best line through the points and use a ruler for straight lines.

Multiple graphs

Sometimes we can show two or more sets of numbers on the same axes. When we put two or more sets of data on the same axes it is easier to draw and compare them. For example, if we want to compare the **temperature** in the school grounds in the sun and in the shade during one day then we could draw two lines on the same graph.

For you to try

1 Look at the information in the table below and draw a double graph on the same set of axes to show the temperature in the sun and in the shade. Use the list of **Notes to help you draw line graphs** and remember to include a key to show the difference between the two sets of numbers.

A

Temperature in school grounds

	Temperature	
Time	in sun (°C)	in shade (°C)
8 am	19	18
9 am	21	20
10 am	23	22
11 am	26	23
12 noon	29	25
1 pm	31	26
2 pm	32	27
3 pm	30	26
4 pm	27	23
5 pm	25	21

2 Show the following information in two bar graphs one above the other. One will be for Port Moresby and the other will be for Lae. Use the same key for both graphs.

Comparison of water use in Port Moresby and Lae

Water use	Percentage used in Port Moresby	Percentage used in Lae
Bathroom	16	22
Toilet	14	19
Laundry	10	13
Kitchen	5	4
Garden	55	42

3 Draw a line graph to show the following data.

Number of wagtails seen on the 15th day of each month

Month	Wagtails	Month	Wagtails	Month	Wagtails	Month	Wagtails
January	23	April	24	July	10	October	14
February	29	May	17	August	9	November	16
March	33	June	12	September	8	December	20

Science in the global environment 

Science can help us to understand the world and to improve life in the community.

People who learn science will have a better understanding of what is going on in the world. This can be very useful when you make choices about the way that you live and when you need to understand information about your environment. Examples of the way in which we make use of science are sometimes called the applications of science.

For example, science can help us to:

- understand how things work, to maintain them and to fix them when they go wrong, for example, pressure lamps, bicycles and engines
- use tools and simple machines to do work, for example, levers, wheels and axles, **inclined planes**, **pulleys** and gear wheels
- use technology in our lives, for example, radios, telephones and computers
- grow and prepare good food and eat a healthy diet—also preserve food so that we can use it later
- prevent sickness and know what to do when we get sick, for example, keeping clean and avoiding smoking and chewing betel nut, avoiding sexually transmitted diseases
- make good decisions about the way that we use the environment, for example, fishing, **logging**, mining and getting rid of waste.

Managing forests

Traditionally people in Papua New Guinea used the forests to make gardens, to get timber to build houses and make canoes and drums, and as a place to find medicines. It was hard work cutting down big trees so they used only what they needed for themselves and did not cut down large areas of forest.

However, after Europeans arrived in Papua New Guinea, sawmills were started and then a logging industry developed to supply timber locally and to other countries. Machines like chainsaws, big trucks and ships make it very easy to cut down large forests and send the logs all over the world. The problem is that tropical rainforests all over the world are now being cut down much faster than they are regrowing or regenerating. In Papua New Guinea the number of trees in the rainforest has been declining very quickly for the same reason. New Ireland has already lost almost all of its rainforest because of poor management. We now need to use our knowledge of science to help make decisions about the logging and timber industry.

In 1988 the Government of Papua New Guinea set up a major enquiry into forests due to the poor management that had been used in the past.

Rainforest in the Hunstein Range, East Sepik

They realised that forest **resources** need to be managed properly otherwise this valuable resource would be lost.

Some of the methods used to control the overuse of forests in PNG are as follows:

1 The total number of trees cut down can be controlled. When all the trees are cut down this is called **clearfelling**. Clearfelling often results in bad soil erosion which makes it impossible for the trees to regenerate or grow again. Using their knowledge of science, ecologists are able to calculate the number of trees that can be cut each year so that the total population of trees remains steady through **regeneration**.

2 Young trees should not be cut down but should be left to grow so that they can be harvested in the future. This is called **selective logging**. For example, in one timber area in the North Solomons, only trees with a diameter between 50 centimetres and 150 centimetres are cut down. This allows the small trees to continue to grow under the protection of the bigger ones that are left.

3 Instead of leaving the natural young trees to grow, an area can be replanted with trees grown in a nursery. This way of managing forests has been used in West New Britain.

In order to make sure that a logged area can grow trees again we must protect the soil because it provides the nutrients for regrowth. The soil must not be packed down or eroded. The heavy machines that are used to drag out the logs must have wide tyres or tracks. These do not pack down the soil as much because the **weight** of the machine is spread over a large area.

We must make sure that steep slopes are not logged because of the risk of erosion. When the eroded material is deposited further down the river it can block the natural drainage channels and cause flooding.

Young kamarere trees in the nursery before being planted out

Logging in Papua New Guinea

It is necessary to keep strict control over logging companies for the following reasons:

- The value of each log depends on the type of tree it comes from, the grade or quality of the log and the size of the log. Some logging companies have been dishonest about the value of the logs they export and say that the logs are worth less than their real value. These logging companies set up a second company overseas and then sell the logs at a lower price than they are worth—so they are really selling the logs to themselves. The second company then sells the logs to a third company at the true value and makes a big profit.
- These methods of cheating are called transfer pricing and it means that logging companies do not pay the correct tax to the government. It is estimated that PNG is losing K200 million each year due to transfer pricing.
- Illegal logging by companies that do not have a licence from the government also means that the environment may be spoilt and money is lost because the companies do not pay taxes.

The National Forest Management Plan

The National Forest Management Plan has been developed by the Department of Forests as a way of managing the use of forests in Papua New Guinea. However, some people think that the plan does not take enough care of the environment and that too many trees will still be cut down.

Forest management in the East Sepik

The Hunstein Range is in the middle of the East Sepik south of the Sepik River. Most of the land is covered with tropical rainforest and the highest land is about 1500 metres above sea level. There is a wide range of ecosystems that have had very little disturbance by human activity and so there are many different plants and animals.

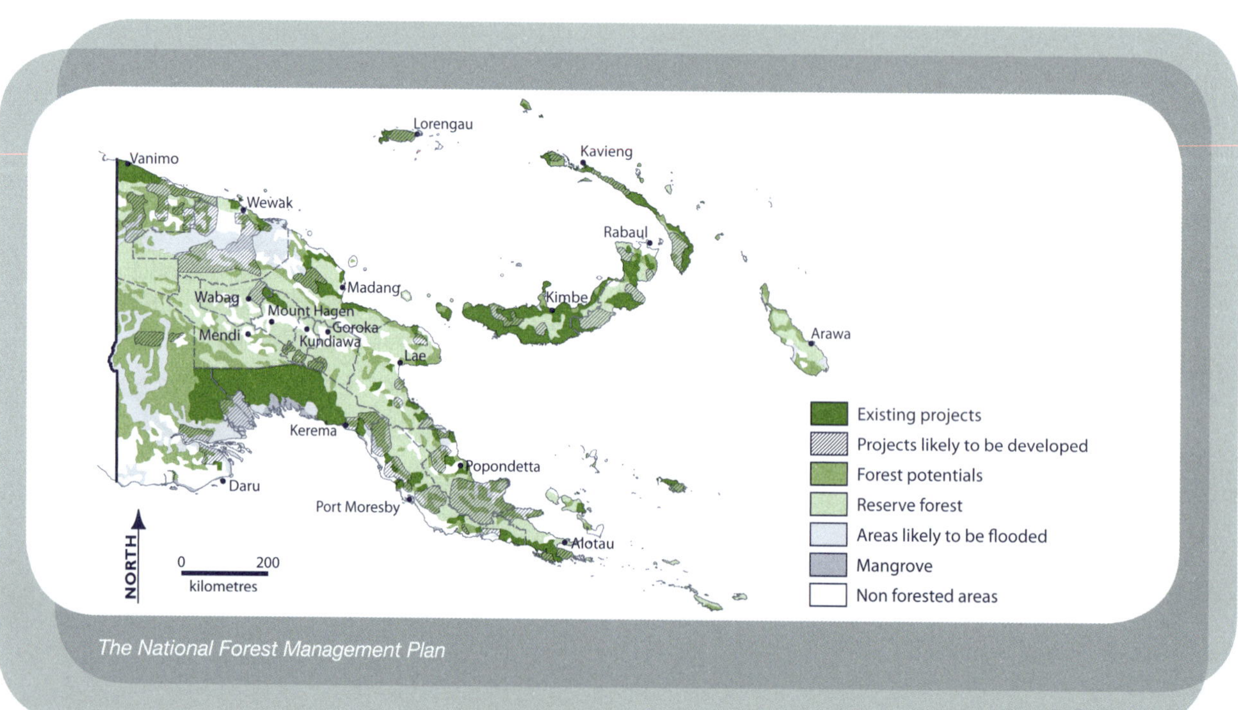

The National Forest Management Plan

The culture and economy of the people is based on the rainforest. They are subsistence farmers—they make sago and also go hunting and collect useful plants from the bush. The way in which the people depend on the forest for their way of life is shown in the diagram below.

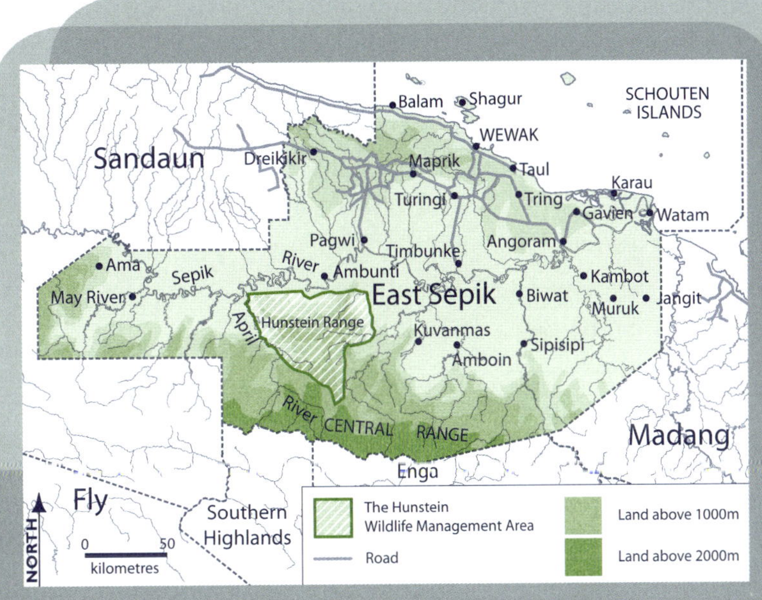

The East Sepik showing the Hunstein Wildlife Management Area

Landform	Sago swamp	Hill forest	Ridge village	Multi-crop garden	River village	River
Vegetation	swamp forest	mixed tropical lowland forest	planted forest and exotic ornamentals and trees	exotic and local economic plants	planted forest and exotic ornamentals and trees	
Food crops	sago, pandanus	lau lau, other fruits, galip nut, greens, e.g. tulip	lau lau	taro, banana, breadfruit, betel nut	lau lau, coconut	fish, crocodiles
Cash sources	sago	cassowary, cockatoo, canoe wood	betel nut	betel nut	betel nut	fish
Other uses	medicine	medicine, fuel, canoes, timber, building materials, dyes, bilum string, water, fish nets	shelter, cooking, living, medicine, kapok, timber, building materials	kapok, timber, building materials	medicine, fuel, canoes, timber, building materials, dyes, bilum string, water, fish nets	water
Cultural use	sago	body paints, bush meat, perfumes, decorations, ceremonies, ritual, story	ceremonies, story, carving, betel nut	kapok, timber, building materials	body paints, bush meat, perfumes, decorations, ceremonies, ritual, story	

Diagram of a cross-section through the Hunstein Wildlife Management Area

The Hunstein people worked with people from the government and non-government organisations to decide what to do. There were two main choices:

- Allow commercial logging which would bring big changes to their way of life but provide money for a while.
- Maintain their environment and way of life by having their area declared a Wildlife Management Area.

In order to help decide the future of their area, the people did the following:

- Made a map of their area and gave a value to all the activities that took place in it—they decided how much money each activity was worth to them.
- Made a list of all the plants and animals that lived in the area.
- Wrote down their traditional stories and culture.
- Made a list of activities that could earn money from the forest but would maintain the environment and their way of life, for example, portable sawmilling, collecting bark that has a perfume, butterfly and orchid farming, carving and making bilums.

After thinking about their choices, the people decided that they would protect their environment and culture by having their area declared as a Wildlife Management Area. This means that commercial logging cannot take place but they can still use the forest to support their way of life.

Managing the seas

The sea is an important resource in Papua New Guinea and in many other parts of the world. The sea provides fish, such as barramundi and reef fish, as well as prawns, lobsters and shellfish. In some places people make clam gardens by taking care of clam shells and allowing them to grow big before harvesting. People also use shells as money, to bail the water out of canoes and to communicate with each other.

We can use our knowledge of science to manage the sea as a resource for the future. For example, when more fish are caught than are born and grow to maturity, this is called **overfishing**. When overfishing takes place, the fish population declines and may eventually die out.

Traditional methods of catching fish did not remove large numbers of fish. People only caught enough for their own needs. However, large commercial fishing boats are now using modern fishing equipment to catch large quantities of fish, squid and prawns. Management must keep strict control over these fishing companies to protect our fisheries for the future.

Methods of management

1 Limit the total number of fish caught. Ecologists are able to calculate the number of fish that can be caught without reducing the population overall. We can use this knowledge to make rules that prevent overfishing.

For you to try

1 Explain what is meant by selective logging and clearfelling.

2 In groups discuss the case of the Hunstein people. Make a list of the advantages and disadvantages for them of living in a Wildlife Management Area. Share your list with other groups. (L) (A)

b) banana prawn

a) barramundi

c) spiny lobster

Three animals from the sea that are important in Papua New Guinea: (a) barramundi (b) banana prawn (c) spiny lobster

For example, catching prawns in the Gulf of Papua is limited by the number of licences that are issued, which stops too many companies fishing. This is necessary because prawns are caught by dragging large nets behind boats. This method is called **trawling** and catches large numbers of prawns. Regulations also control the number of nets used on each boat and the size of the boats.

A short drift net that does not catch too many fish

2 Limit the size of fish that are caught. Young fish should be left as breeding **stock**. Older fish can be caught selectively by using nets with a larger mesh size that allows the smaller, younger fish to swim through. For example, a minimum gill net size of 7–8 centimetres has been introduced west of Daru because young barramundi breed there. After breeding, barramundi migrate up the Fly River and other inland rivers. A year later they return to the sea to breed again.

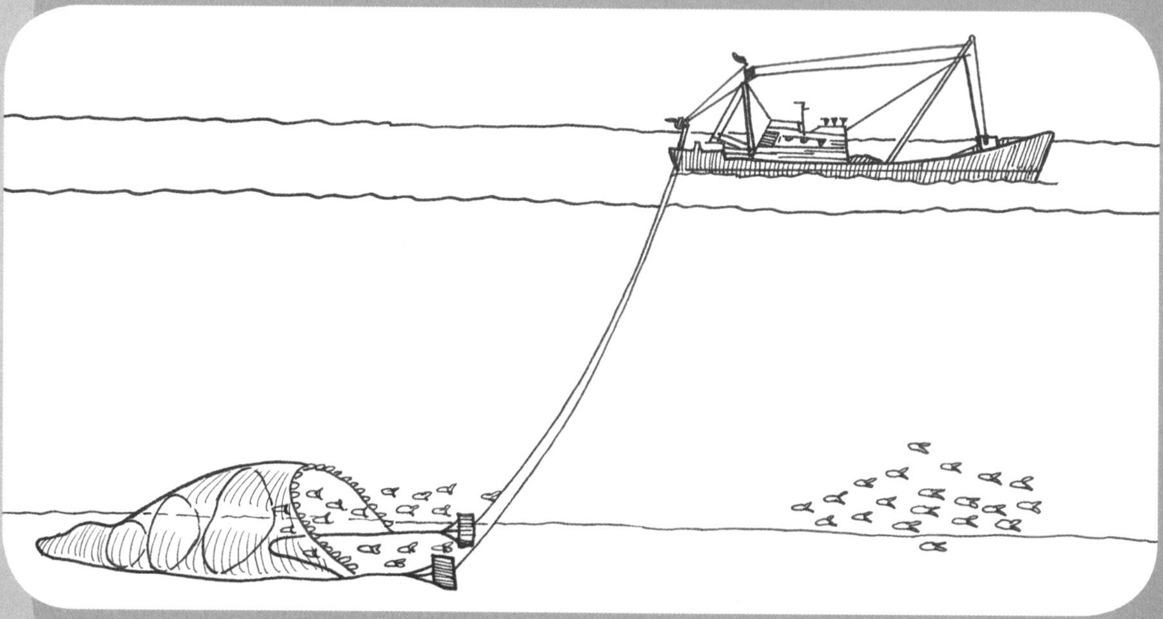

A trawl net being pulled behind a trawler

The main breeding grounds and migration route of barramundi

3 Limit the season when fishing is allowed. Fishing should not be allowed when the fish are breeding or migrating. Catching fish at these times will stop the **life cycle** of the fish, which results in fewer fish being produced.

Breeding grounds and migratory routes need special protection from disturbances and **pollution** to make sure that the maximum number of fish can be born. Ecologists think that lobster trawling in the Torres Strait has disturbed the migration and resulted in very few lobsters in the Gulf of Papua. As a result trawling in the Gulf of Papua has been banned.

Prawn trawling is also banned within three kilometres of the coast to protect the breeding grounds of the prawns. Fish also need a plentiful supply of food to breed successfully.

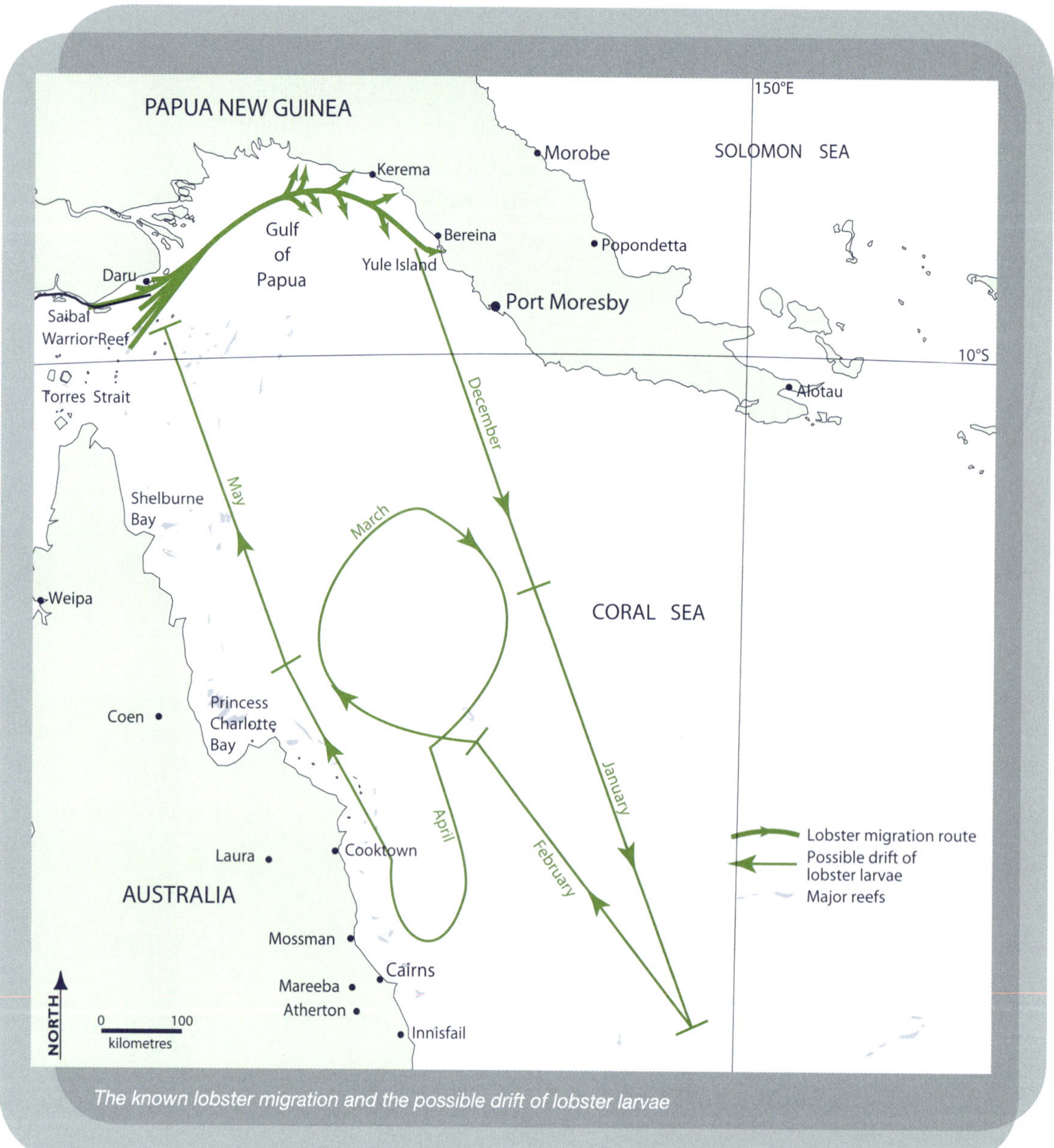

The known lobster migration and the possible drift of lobster larvae

Ecologists can study the food web that the fish is part of and help to protect its food sources. For example, mangroves are an important place for young fish to grow and live.

4 Limit the use of drift nets.
Some countries fishing in the South Pacific are using **drift nets** as long as 50 kilometres. These nets are called drift nets because they are allowed to drift in the sea. These nets catch all kinds of fish, as well as dolphins, dugongs, turtles and even sea birds. They are also called the 'wall of death' because they kill so many different animals. Ecologists are worried that the use of drift nets will result in the overfishing of tuna as well as seriously affect other animals in the sea.

For you to try

1 Describe three methods of managing the seas.

L A

2 In groups discuss why it is necessary to manage natural environments such as the forests and the sea. Summarise your main ideas and share these with other groups.

SS

Projects

Using the list on page 16 under 'Science in the global environment', identify a project that could be carried out in your area. The project should make use of the knowledge and skills that we get from science. Plan your project and carry it out in your school or community. Some ideas are given below:

- Find out about different ways that food is preserved in the community such as drying, smoking, salting and pickling. Carry out these methods on different kinds of food and compare the results. Which methods can easily be used in the community?
- Find out about the main health problems in your area by carrying out a survey. Choose one of the problems and try to find ways to help solve the problem. For example, ways to prevent sickness and knowing what to do when we get sick, such as keeping clean and avoiding smoking and chewing betel nut, and avoiding sexually transmitted diseases.
- Carry out a survey to check some part of the local environment such as the side of the road, a beach, river or creek. Try to identify any problems and their cause. Decide how the problems can be cleaned up and prevented from happening in the future. If possible, carry out your plans.

Summary questions

1 What sense would you mainly use to detect the following:
   a the temperature of bath water?
   b sugar in a cup of tea?
   c a key that has been dropped in the dark?
   d an outboard motor when you are spear fishing underwater?
   e a small animal that has crawled out of sight and has died?

2 Answer true or false:

When an experiment is fair …	T or F?
somebody else should also be able to repeat the experiment and get similar results	
getting different results when the experiment is repeated will help us to feel more confident about our conclusions	
repeating the experiment can help to fully understand the reasons why something is happening	
we will always see a pattern in the results	

3 Which of the following best describes the sequence of steps that we follow when we carry out a scientific investigation?
   I Choose suitable methods that are fair.
   II Make conclusions relating to the question.
   III Ask a suitable question.
   IV Get some results.
      A III, I, IV, II
      B III, IV, I, II
      C III, I, II, IV
      D II, IV, I, III

4 Which of the following best describes the order of the questions that we ask when we carry out a scientific investigation?
   I What happened?
   II What do you want to find out?
   III What do the results mean?
   IV How will you find out?
      A III, IV, I, II
      B II, IV, III, I
      C II, IV, I, III
      D III, I, IV, II

5 Look at the following statements about the way that we use science and answer true or false:

Statement	T or F?
Science can help us to understand the world and improve life in the community.	
People who learn science are less likely to understand what is going on in the world than those who do not learn science.	
Our knowledge of science is useful when we make choices about the way that we live and need to understand information about the environment.	
The applications of science are examples of the ways in which we make use of science in our everyday lives.	

6 Which of the following are good examples of managing the forest?

- **I** clearfelling
- **II** selective logging
- **III** replanting
  - **A** I and II only
  - **B** II and III only
  - **C** I and III only
  - **D** I, II and III

7 Which of the following are good examples of managing the seas?

- **I** Limit the total number of fish caught
- **II** Limit the size of fish that are caught
- **III** Limit the season that fish are caught in
- **IV** Limit the use of drift nets
  - **A** I, II and III only
  - **B** II, III and IV only
  - **C** I, II and IV only
  - **D** I, II, III and IV

8 When we compare managing the forest with managing the seas, then clearfelling is most similar to

- **A** trawling
- **B** overfishing
- **C** using drift nets
- **D** fishing during the season when it is not allowed

9 When we compare managing the forest with managing the seas, then selective logging is most similar to limiting the

- **A** use of drift nets
- **B** size of fish that are caught
- **C** total number of fish caught
- **D** season when fishing is allowed

10 The passage opposite is a summary of the main ideas of this chapter. Copy and complete the passage in your book. Using the words in the list, find the words that are missing. You can use each word only once.

accurate, controls, decisions, fair, information, investigations, method, purpose, questions, sensible, understand

People want to understand the world around them and find out the answers to many different ____________. A good way to find more about why things happen and how they work is to carry out scientific ____________, which usually means that we try to find the answer to a question. When we carry out an investigation we must make sure that we are asking a question that will help us to ____________ and then choose a way to find an answer to that question. When we carry out investigations we must also try to make sure that we are being ____________, and this usually means that we must use ____________. Each investigation will be different because it has a different ____________, but we must always be ____________ in the way we collect and record information. The conclusions that we make must be based on the ____________ that we have collected and be linked to the purpose of the investigation. When we think that we have found an answer we must look at the question again and decide if the answer is ____________ or reasonable. When we use the scientific ____________ to collect reliable information and answer questions we can make good ____________ about the way that we can use science in everyday life.

2

Living things

Chapter summary

In this chapter you will have an opportunity to:

- find out about the life cycle of living things
- find out about stages of reproduction in living things and draw some of these stages
- collect and use information to compare reproduction in different living things from various habitats
- find out about the relationship between the environment and the processes that living things carry out
- make your own test to show the difference between biodegradable and non-biodegradable substances
- help other people in the community to understand about the wise use of non-biodegradable materials.

Syllabus references

Strand: Living things

Sub-strand: Nature of living things
Ecology, Relationships and Interactions

Outcomes:

8.2.1 Describe and explain processes of reproduction in living things and how the environment influences these processes.

8.2.2 Draw conclusions regarding the effects of excessive use of non-biodegradable materials on food webs.

Key facts

- All living things pass through a life cycle which consists of different stages.
- At some stage in the life cycle, living things are ready to reproduce.
- Reproduction is necessary for the species to survive, but an individual plant or animal does not need to reproduce in order to complete its own life cycle.
- Reproduction can be asexual or sexual.
- In asexual reproduction the new individual is produced from one parent only. There are different types of asexual reproduction.
- In sexual reproduction the new individual is produced from two parents.
- In sexual reproduction the male sex cell and the female sex cell join together during the process of fertilisation.
- Fertilisation that occurs outside the body of the female is called external fertilisation and fertilisation that occurs inside the body of the female is called internal fertilisation.
- Different living things reproduce in different ways depending on the environment in which they live.
- All living things will eventually die then rot or decay.
- Materials that do rot or decay naturally are called biodegradable materials.
- Materials that do not rot or decay naturally are called non-biodegradable materials.
- We need to use all materials wisely, and understand how to get rid of non-biodegradable materials that are no longer useful.

Reproduction (PD)

People who lived hundreds of years ago believed that living things came from non-living things. For example, when it rained after a long dry season they saw frogs appearing from the mud. They also saw flies appearing from decaying meat. The living things just seemed to appear by themselves and so this theory was called **spontaneous generation.**

In 1668 an Italian doctor called Francesco Redi carried out an experiment to test the theory of spontaneous generation. He set up three jars each containing some meat. Jar 1 was open to the air, Jar 2 was covered with gauze that allowed air inside and Jar 3 was covered with parchment that did not allow air inside.

After several days he noticed that maggots appeared on the meat in the open jar but did not appear on the meat in the other two jars even though the meat was rotten in all three jars. However, maggots did appear on top of the gauze in Jar 2 where the flies had landed. He concluded that the maggots did not come from the decaying meat or from the air but must have come from the flies. However, Redi still believed that spontaneous generation occurred in some cases such as worms that live inside the gut of some animals.

It was not until 1878 that the theory of spontaneous generation was shown to be completely wrong. A French man named Louis Pasteur believed that germs could be found on all the surfaces of all objects, in the air and in the water. He showed that these germs that cause things to become rotten can be killed by heating them to 55°C. This simple process in now known as **pasteurisation** and was first used in making beer, but is now also used to make milk and other foods safe to eat and drink. Another advantage of pasteurisation is that it does not change the taste of the food.

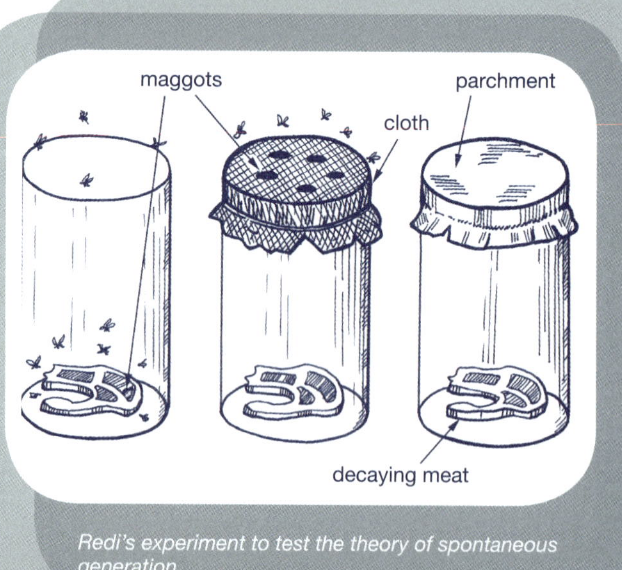

Redi's experiment to test the theory of spontaneous generation

For you to try

1 Describe two pieces of evidence that helped show that the theory of spontaneous generation was not correct. **L** **A**

2 Redi's experiment was one of the first to use a test and proper controls. Explain which jar was the test and which were the controls. Also explain why it was a fair experiment. **L**

The need to reproduce

All living things pass through a life cycle that consists of different stages. At some stage in the life cycle, living things are ready to reproduce. Reproduction is needed for the species to survive since all living things die at the end of their life cycle. Reproduction is one of the characteristics of all living things but it is not necessary for an individual plant or animal to reproduce in order to survive and complete its own life cycle.

Different species reproduce in different ways, depending on the environment in which they live. This is called an adaptation to the environment. For example, frogs lay **eggs** in water or damp places while **reptiles** lay eggs that have shells on the land. Frogs' eggs cannot survive without water. When a female dog is on heat she is ready to mate and produces a smell that attracts other dogs near where she lives. This is also an adaptation.

There are two main types of reproduction, **asexual reproduction** and **sexual reproduction**.

Asexual reproduction

When a new individual is produced from one parent only, the process is called asexual reproduction. There are five main types of asexual reproduction.

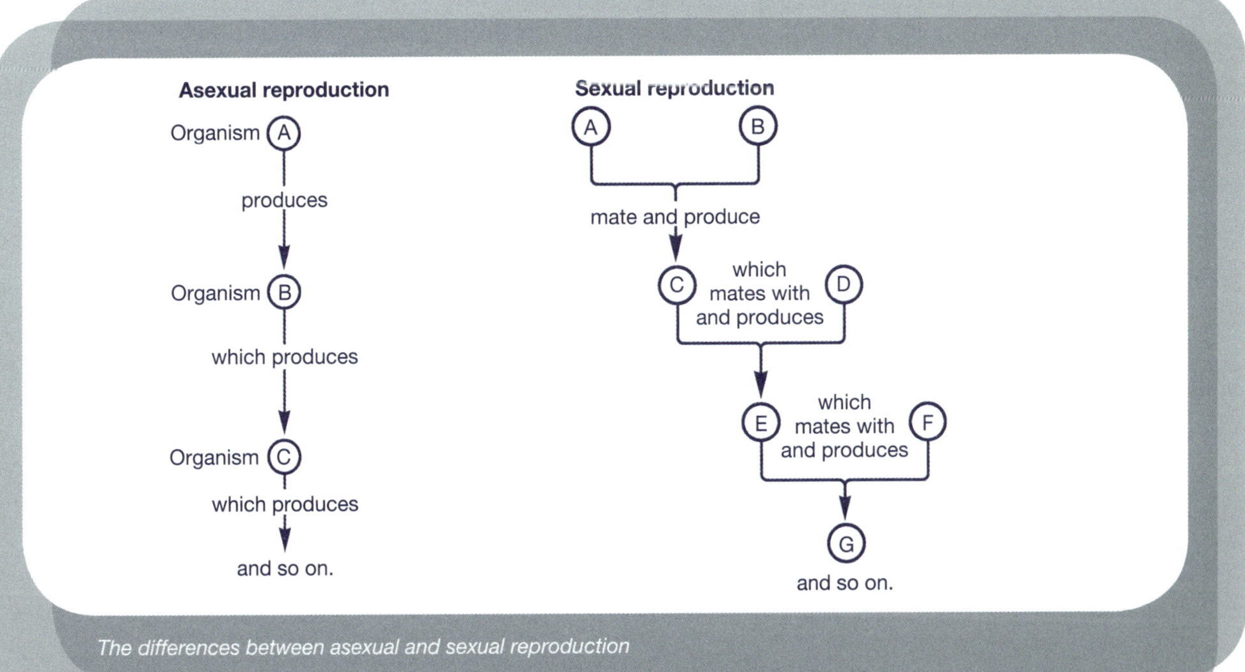

The differences between asexual and sexual reproduction

1 Simple or binary fission

Simple fission, also known as **binary fission**, occurs when one cell divides into two cells and the contents of the two cells are shared. This kind of reproduction is found in simple living things such as bacteria, protozoa and **algae**. This method allows some bacteria to reproduce every twenty minutes, which can cause food to go bad very quickly or make us sick in a very short time.

2 Spores

Spores are single cells that are very light and usually produced in large numbers. **Fungi**, **mosses**, **ferns** and algae produce spores. For example, ferns produce spores in special structures on the underside of their 'leaves' or fronds. These structures can be green, black or brown and are a little rough so the fern looks like it has a disease even though it is normal. Mushrooms also produce thousands of spores from their gills, which are under the cap.

Spores develop into new organisms when they find suitable conditions.

3 Budding

Some organisms reproduce by forming a bud that slowly develops into a new individual. This is called **budding**. Later it may separate from the parent and live as another adult. For example, yeast that is used to make bread, beer and wine is able to reproduce like this.

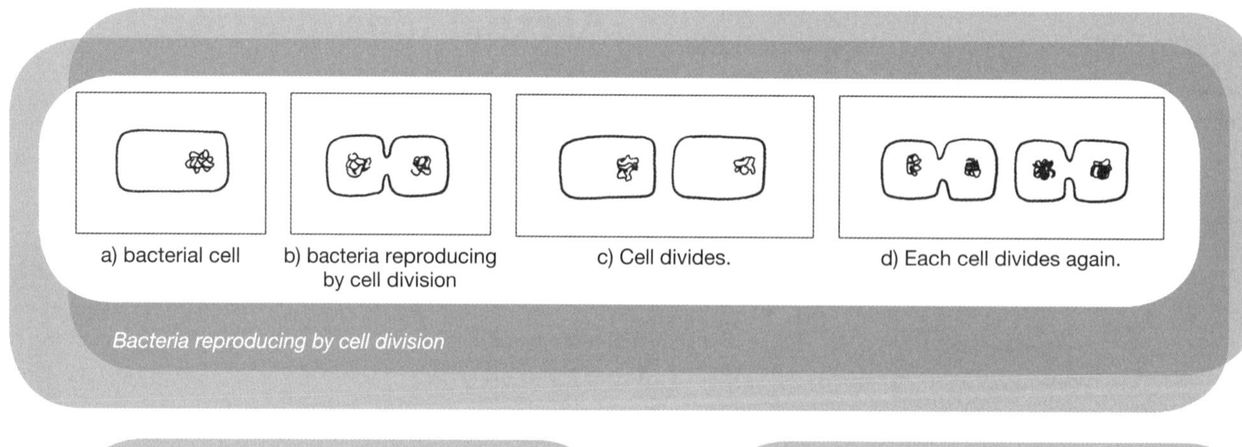

Bacteria reproducing by cell division

Spores are produced on the underside of the leaves

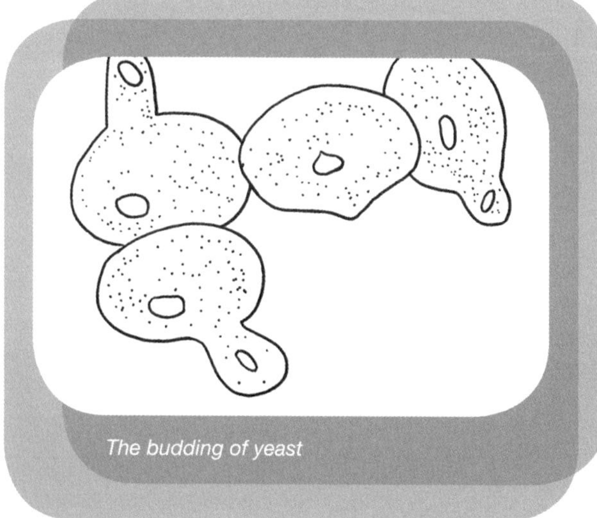

The budding of yeast

4 Regeneration

In some animals, new individuals can develop from pieces of the adult. This is called regeneration and occurs in sponges and sea stars. For example, a new sea star will grow from a piece of the adult as long as a part of the mouth tissue is present.

5 Vegetative reproduction

Vegetative reproduction occurs in plants when a new individual is produced from any part except the flower. When we plant cuttings in the garden we are making use of vegetative reproduction. Examples of vegetative reproduction are shown in the diagram below.

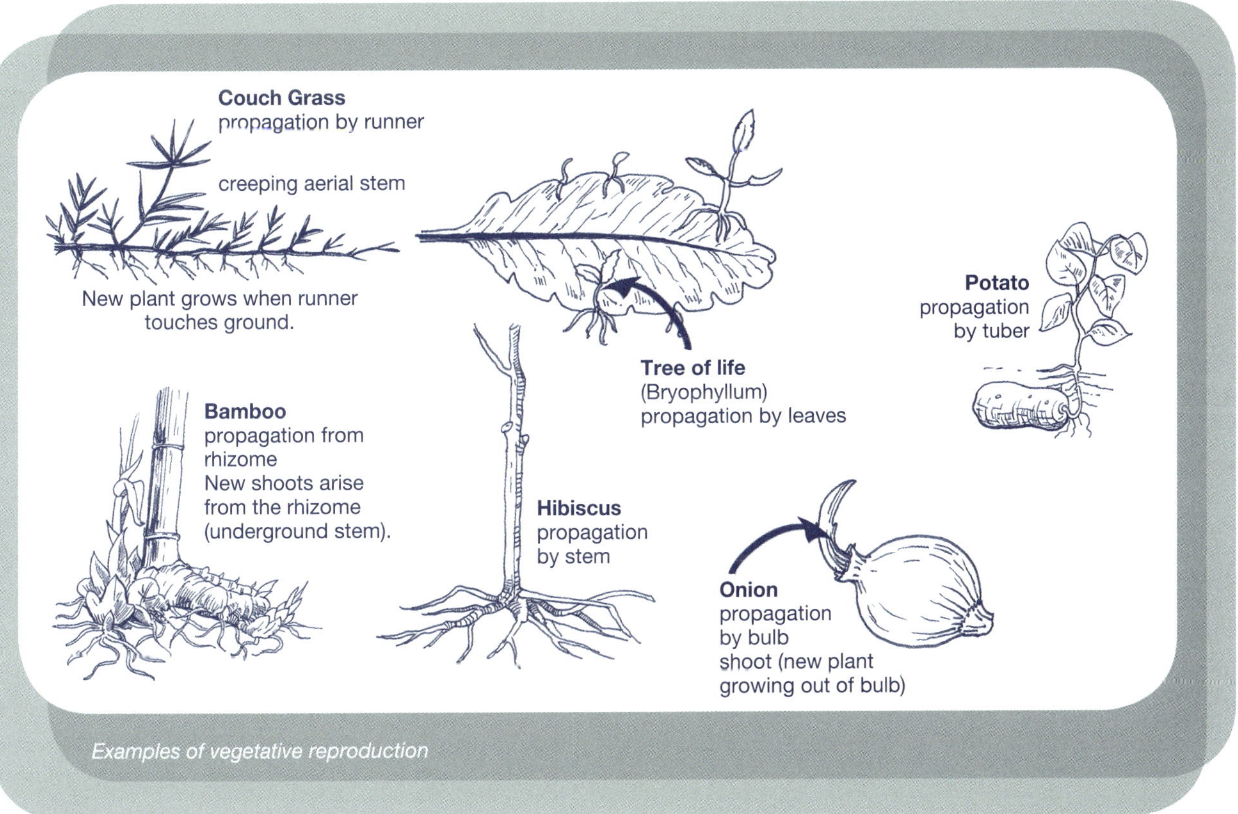

Examples of vegetative reproduction

For you to try

1 **Investigation: Looking at fern fronds**

a Collect a number of different fern fronds and a magnifying glass.

b Carefully examine the underside of different fronds and look for the special structures that contain the spores.

c Make a clear labelled drawing of what you observe.

2 **Investigation: Making a mushroom spore print**

a Collect a mushroom and a sheet of plain paper. (continued over) b Carefully remove the stem and place the cap on a clean sheet of paper with the underside against the paper.

c Leave the mushroom for several days and make sure it is not moved or disturbed.

d Carefully remove the mushroom cap from the paper and observe what has happened. This is known as a spore print.

e Give a title to the piece of paper, add some labels and information about where the mushroom was found.

f Explain what you observe on the paper.

g Compare different spore prints if you can.

3 What are spores? Why do plants like ferns produce thousands of spores rather than a few spores?

4 Imagine that one cell of a bacterium can reproduce every twenty minutes. How many cells will there be

a after one hour?

b after two hours?

c after four hours?

d after six hours?

Sexual reproduction

- In both plants and animals, young produced from two parents is called sexual reproduction.
- The male parent produces male sex cells, which are called **sperm** in animals and **pollen** in **flowering plants**.
- The female parent produces female sex cells which are called eggs. The egg is sometimes called an **ovule** in plants.
- **Fertilisation** is the joining of a male sex cell with a female sex cell. This forms a single cell called a **zygote**.

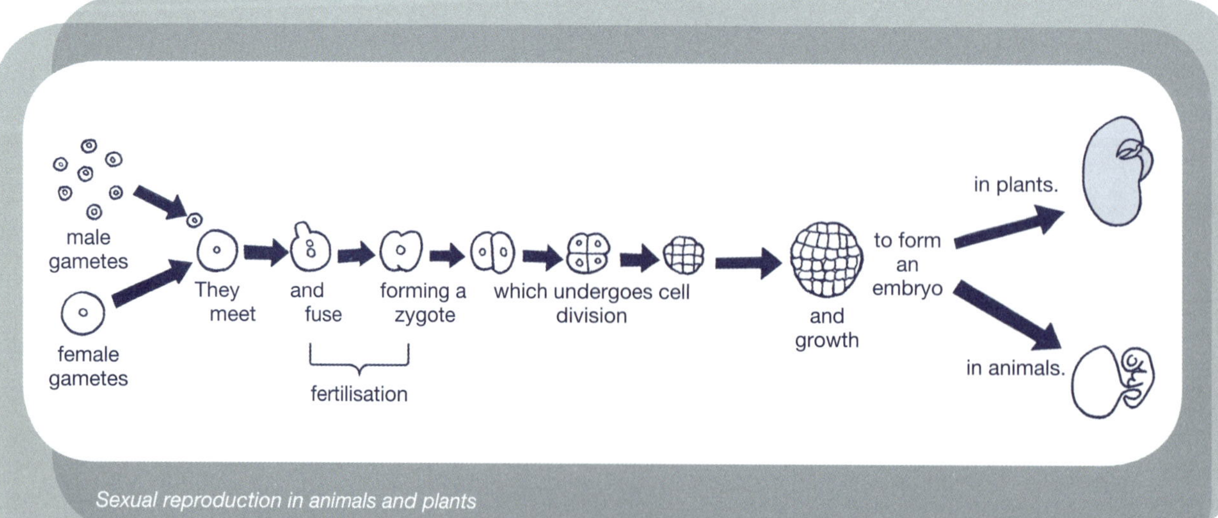

Sexual reproduction in animals and plants

- The zygote then divides and grows to become an **embryo** in both plants and animals.
- In animals there are two main methods of fertilisation: **external fertilisation** and **internal fertilisation**.

External fertilisation

External fertilisation occurs when the two sex cells join together outside the body of the female. For example, in fish and many frogs the female lays eggs in the water and the male then puts his sperm in the water. The sperm are carried to the egg by the water. However, many eggs and sperm are wasted and so large numbers of eggs must be produced in order to make sure that enough will be fertilised.

Internal fertilisation

Internal fertilisation occurs inside the body of the female. Most reptiles, birds and **mammals** use internal fertilisation. In mammals, the **penis** of the male leaves the sperm inside the female which increases the chances of fertilisation and so fewer eggs are produced by these animals.

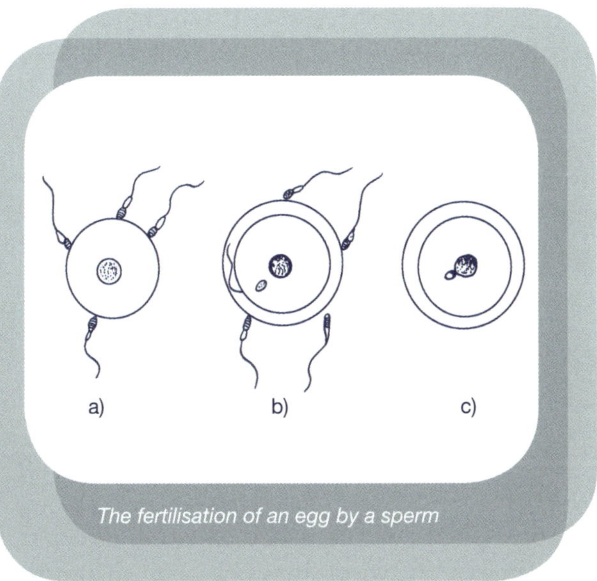

The fertilisation of an egg by a sperm

- The fertilised egg is a single cell called a zygote.
- The zygote divides thousands of times and grows to become an embryo.
- In humans, from the third month of development until birth, the developing baby is called a **foetus**.

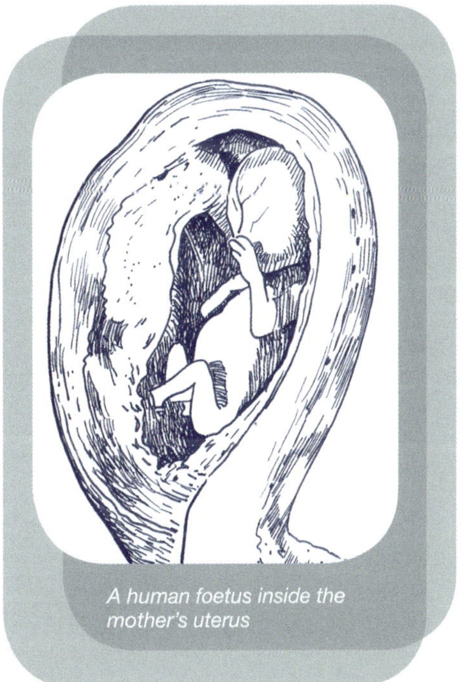
A human foetus inside the mother's uterus

For you to try

1. Give one reason why mammals produce only a few eggs.
2. In human beings, what is the difference between a foetus, an embryo and a zygote?
3. Use some examples to describe some advantages and disadvantages of internal and external fertilisation. A
4. Reproduction is one of the characteristics of all living things, but not all living things reproduce. Explain how this can be possible. L

Reproducing with flowers

The reason that flowering plants have flowers is so that they can reproduce. Flowers contain tiny male and female cells that grow into seeds when they join together.

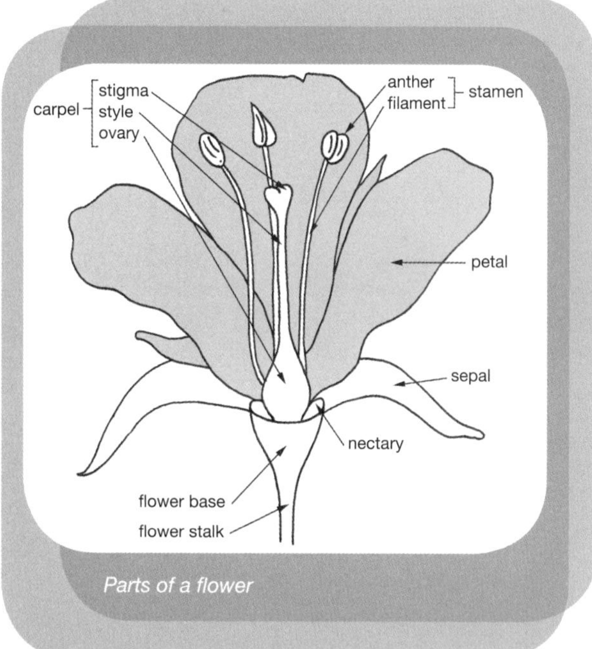

Parts of a flower

Some flowers contain male sex cells and some contain female sex cells. For example, pawpaw trees have separate male and female flowers. However, many flowers contain both types. The main parts of a flower are as follows:

- **Stamens** or male parts that hold the grains of pollen that contain the male sex cells.
- Pollen is released when the **anther** becomes ripe and splits open.
- The **carpels** or female parts that include the ovary, style and stigma.
- In the ovary are tiny ovules each containing a female sex cell.
- When the tip of the stigma is ripe it is ready to receive grains of pollen.

Pollination

Flowering plants are not able to move around like animals and so there must be another way to bring the egg and the pollen together. Pollen is usually carried from one flower to another flower by wind, water or an animal. This process is called **pollination**.

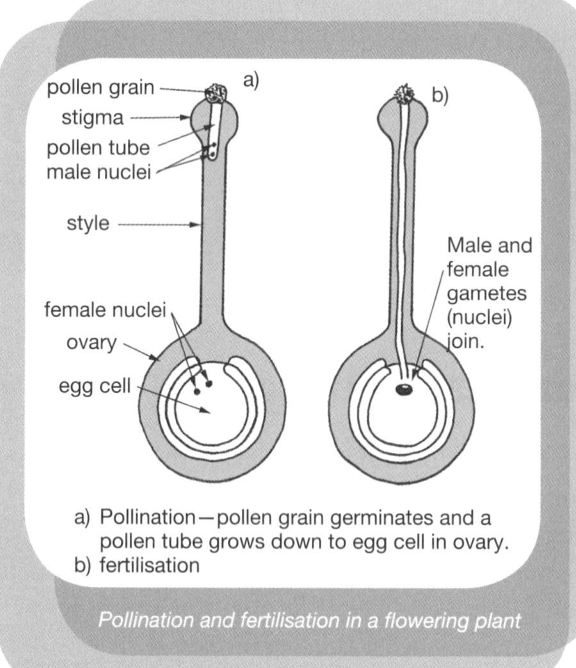

Pollination and fertilisation in a flowering plant

Self-pollination means that pollen is carried from stamens to carpels on the same plant.

Cross-pollination means that pollen is carried from stamens to carpels on another plant of the same type. There are different ways in which this can happen:

- Some plants are pollinated by insects. Insects such as bees carry pollen on their bodies as they move from flower to flower. The insects are attracted by the flower's smell or bright colour. They look for the sugary nectar inside the flower.
- Some plants are pollinated by wind. Pollen is carried by the wind from one plant to another.

The flowers are not usually brightly coloured, but have parts which hang out in the wind.

Cross-pollination is useful because it gives a greater variety of young plants than self-pollination. Some of these plants will survive better than others, which will help the species to be successful. Flowers often have stamens and carpels that ripen at different times which prevents self-pollination.

a) cross-pollination between flowers on different plants

b) cross-pollination between different flowers on the same plant

c) self-pollination

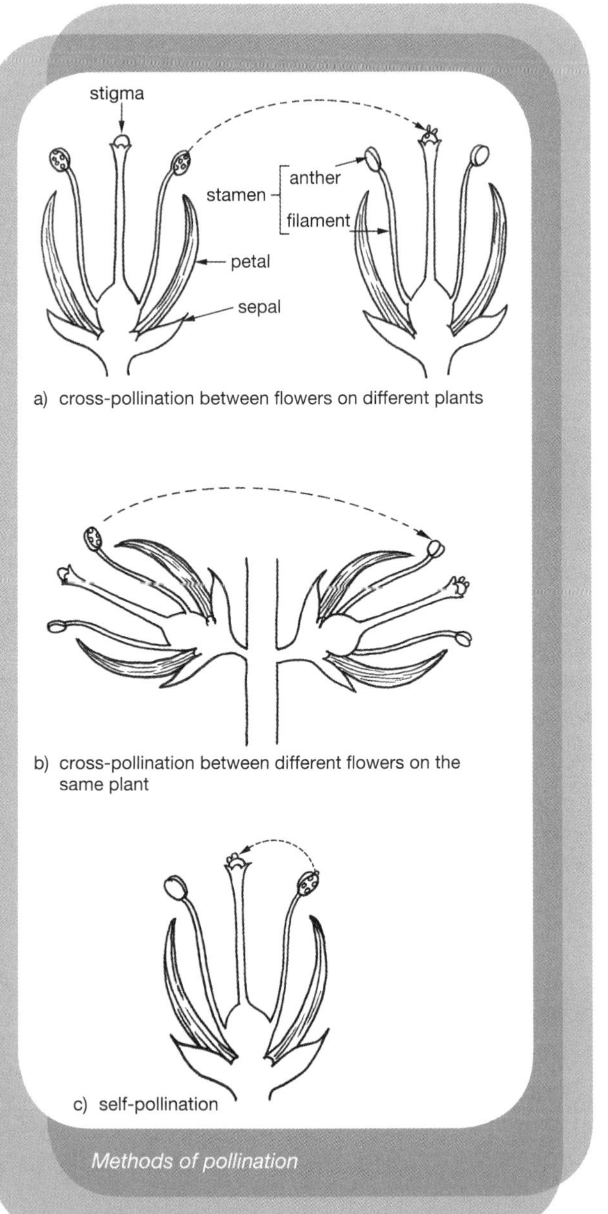

Methods of pollination

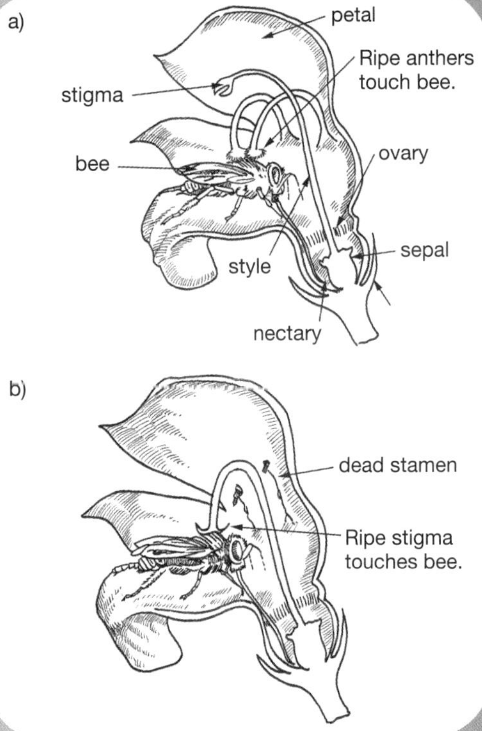

Cross-pollination in a typical flower
In a) the anthers are ripe but the stigma is unripe. Pollen is dusted on to the bee.
In b) the stigma is ripe, but the anthers are dead. Pollen rubs off the bee onto the stigma.

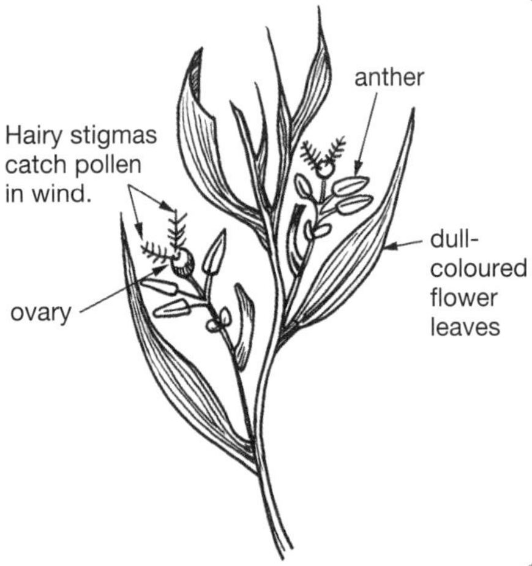

Wind-pollination in grasses

Fertilisation

When a pollen grain sticks to a ripe carpel, a pollen tube may grow out of the grain and down to an ovule. A nucleus from a male sex cell can then pass down this tube and join with the nucleus of the female sex cell. When this happens the cell has been fertilised.

Seeds and fruits

After fertilisation, a complete ovary is called a **fruit**. In the ovary each fertilised cell grows by cell division to form a seed. Each seed contains an embryo, a food store and a thick coat for protection.

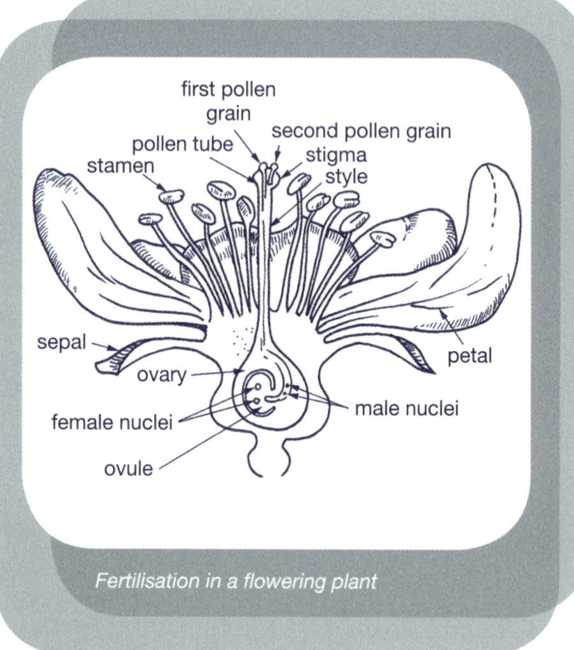

Fertilisation in a flowering plant

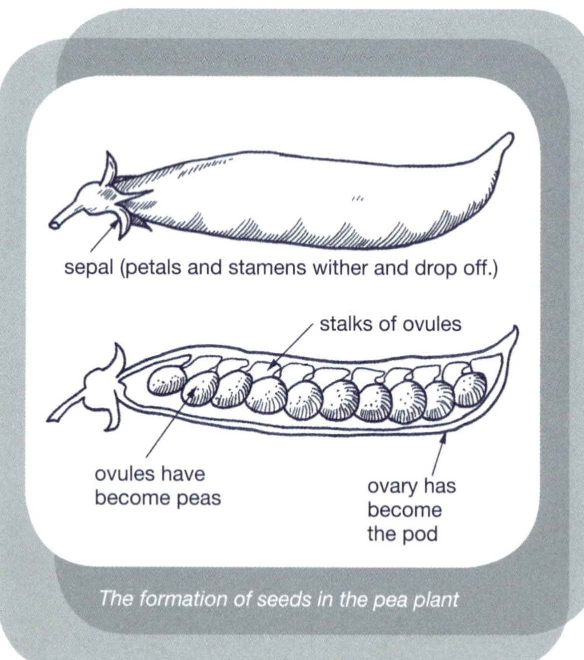

The formation of seeds in the pea plant

For you to try

1 **Investigation: Looking at flowers**
   - a Collect a large flower, a razor blade and a magnifying glass.
   - b Very carefully cut the flower in half by cutting through the stem and then on through the flower. Take extra care since the razor blade is very sharp.
   - c Identify the different parts of the flower.
   - d Draw a labelled diagram of what you observe. Under each label write the function of that part.

2 What is the difference between pollination and fertilisation?

3 Using the names of examples of different flowering plants, explain the different ways that they can be pollinated.

Types of fruit

The shape of the fruit varies from one type of flowering plant to another. Fruits can be put into groups or classified as shown in the table below:

Types of fruit

Name	Characteristics	Examples
Succulent or juicy fruits	part of the ovary becomes juicy or fleshy	mangoes, tomatoes, bananas, melons and cucumbers
Dry fruits—two types:	ripen without pulp or flesh	
• splitting fruits	fruits remain attached to parents but split open to release seeds	bauhinia, peas, beans, flamboyant and kapok trees
• non-splitting fruits	fruits separate from parent but do not split open	okari, corn, cashew, and sunflower

Fruits that can be eaten are called **edible** fruits. Fruits that cannot be eaten are called **inedible** fruits.

Seed dispersal

Plants try to spread or scatter their seeds over a large area so that some may survive and grow into new plants. This reduces the competition for water, sunlight and **minerals**. The scattering of seeds is called **dispersal**. There are four main methods of dispersal and the examples show the way that the plant is adapted to the environment.

1 Animals

Many animals eat fruit and the seeds pass through the digestive system unharmed. The seeds are then left in their droppings a long way from the parent plant. Examples are tomato and passionfruit. Humans also disperse seeds when they harvest fruits and then carry them away to eat them in different places.

Some seeds are sticky and birds may disperse these seeds by wiping them off on a tree branch. The seeds may then fall to the ground and germinate.

Some seeds have hooks or **burrs** and stick to the fur of animals where they can be carried for long distances. Humans can also spread seeds in this way. For example, mimosa and some grass seeds.

People also carry seeds from one country to another using modern methods of transport. However, there are special rules to make sure that plant and animal diseases are not carried from country to country. These rules are called **quarantine**.

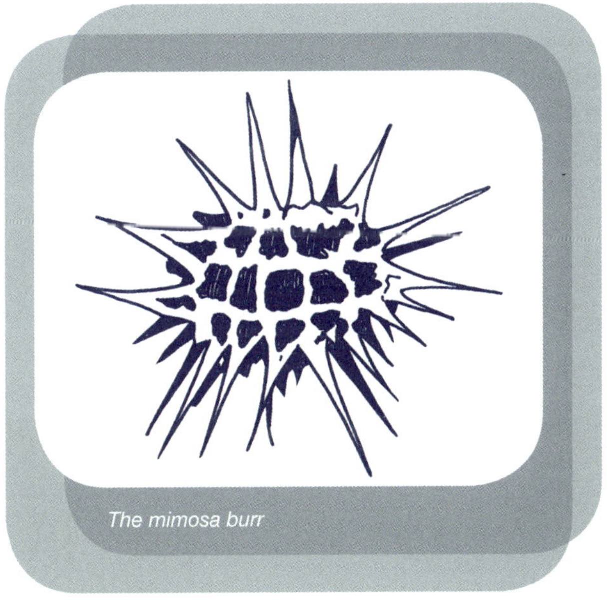

The mimosa burr

2 Wind

Some seeds have wings or parachutes to help them in dispersal. These seeds are usually light and can be carried long distances by the wind. For example, pines, she-oaks and jacarandas have

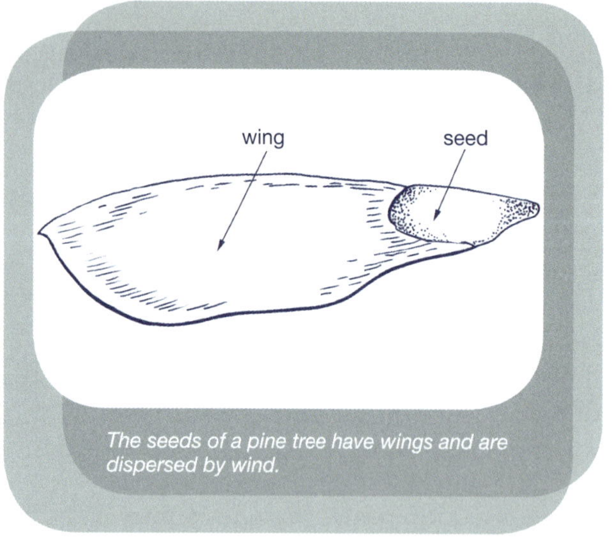

The seeds of a pine tree have wings and are dispersed by wind.

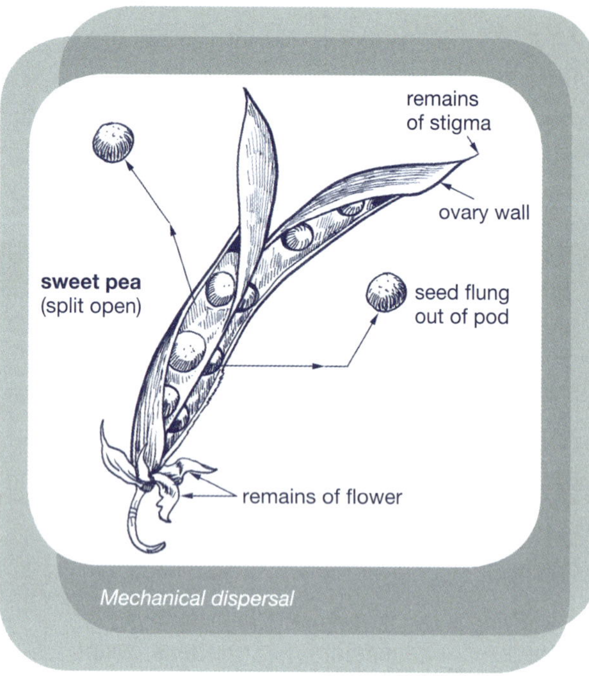

Mechanical dispersal

winged seeds. The African tulip and cotton seeds have parachutes.

3 Water

Some fruits and seeds can be dispersed by water. For example, coconuts are large, **waterproof** fruits that float and can travel large distances on the sea. The seeds of mangrove trees also float and are dispersed by water, and the seeds of sea grasses are dispersed by ocean currents.

4 Mechanical

When some fruits dry out they explode or burst open with such **force** that they scatter the seeds. On a hot, dry day you can often hear the sound of the explosion. Examples are the sweet pea and wild lupin.

Other fruits and seeds may roll a short distance after falling from the parent plant.

Seeds are living although they may remain resting or **dormant** for months or even years. Dormant seeds will germinate when the conditions of water, temperature and air are suitable.

For you to try

1 **Investigation: Seed dispersal**

   a Collect a variety of different seeds and try to work out how far the seeds have moved from the parent plant.

   b Carry out some tests on the seeds with wings or parachutes. Throw them in the air and measure how far they can be dispersed on the wind.

   c Put the seeds into groups depending on the way that they are dispersed.

   d Make a poster by sticking examples of each type of seed on a sheet of paper.

e Label your poster to show the names of the seeds, the way in which they are dispersed and the distance they have been dispersed.

2 What are the four main methods of seed dispersal? A

3 Why is seed dispersal necessary for most flowering plants? A

4 Explain why the method of seed dispersal used by flowering plants is described as being an adaptation to the environment. L

5 Using information in the table Types of fruit explain why tomatoes and cucumbers can be classified as fruits. What characteristics do we normally use to classify fruits and vegetables?

6 Explain why Papua New Guinea has quarantine rules about bringing seeds into the country. L

Reproduction in insects

- Most insects are either male or female.
- The male produces sperm and the female produces egg cells.
- The male and female must join together, usually by the end of the abdomen.
- The male passes sperm into the female to fertilise the eggs.
- The female usually lays eggs on or near a suitable food source. Some female insects have a special tube on the abdomen called an ovipositor for laying eggs.

Green vegetable bugs mating. The male is on the left.

Growth and development

As living things grow and develop from an egg to a mature adult they go through many different stages which are known as the life cycle. In insects, there are three types of life cycle:

1 No-change cycle
2 Part-change cycle
3 Complete change

No-change cycle

- The adults lay eggs which hatch into young that look like the adult.
- A young insect is called a **larva**. Each larva grows until it is too big for its outside covering which is called the **exoskeleton**.

A female cricket, showing the long ovipositor at the rear

- It then loses the exoskeleton in a process called moulting.
- Silverfish and springtails have this kind of cycle.

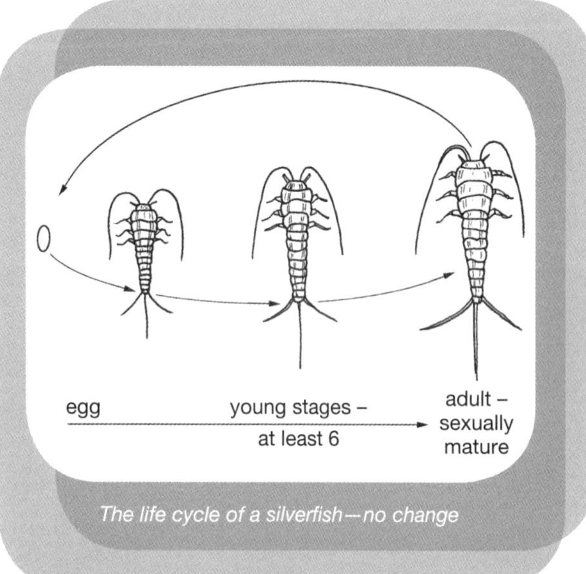

The life cycle of a silverfish—no change

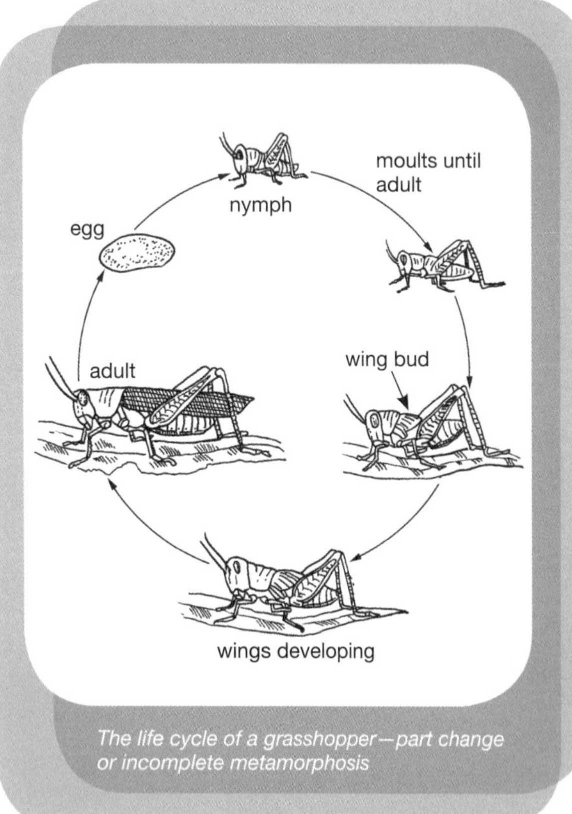

The life cycle of a grasshopper—part change or incomplete metamorphosis

Part-change cycle or incomplete metamorphosis

- There are three stages in this life cycle: egg, larva or **nymph** and adult.
- The egg hatches into a larva which is called a nymph and looks very much like the adult.
- The nymph then slowly changes as it grows and wings develop from wing buds.
- It moults several times to become an adult.
- This type of life cycle is called incomplete **metamorphosis**.
- Grasshoppers, cockroaches, earwigs, termites, dragonflies, praying mantis, crickets and true bugs have this kind of life cycle.

Complete change cycle or complete metamorphosis

- There are four separate stages in this life cycle: egg, larva, **pupa** and adult.
- The egg is laid usually on one type of plant.
- The egg hatches into a small caterpillar or larva which spends most of its time feeding. It grows steadily and moults each time it becomes too big for its exoskeleton.
- At the end of the larval stage many insects spin a special case called a **cocoon** and fix themselves to a twig.
- The third stage of the life cycle is the pupa, which does not feed at all.
- Inside the pupa, big changes take place as the larva changes into an adult.
- It develops wings, a mouth tube, eyes, legs, and many other parts.
- Then the case splits open and the adult or **imago** comes out.
- The adult feeds until it is ready to reproduce.
- Most insects have this kind of life cycle, for example, butterflies, moths, bees, wasps, flies, mosquitoes and beetles.

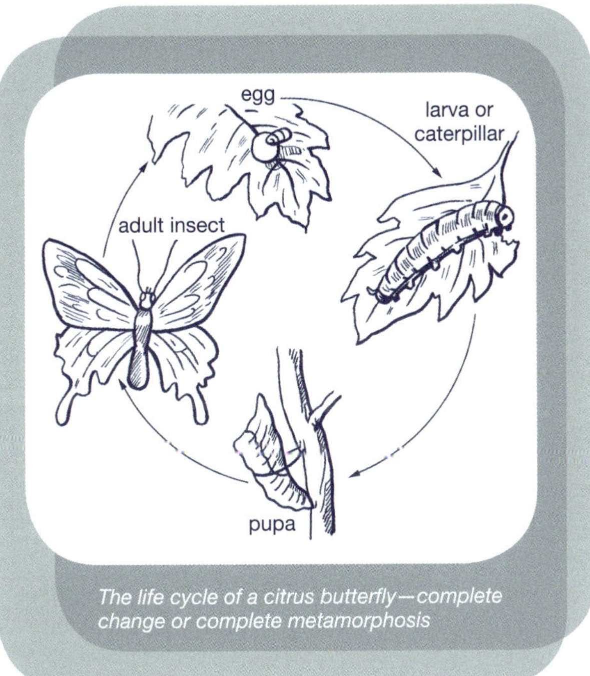

The life cycle of a citrus butterfly—complete change or complete metamorphosis

Adaptations to the environment

Insects are very successful and about three quarters of all the animals in the world are insects. This means that insects are very well adapted to living in different environments. For example:

- Most bees lay their eggs in hives where they live as a group. The pupa develops inside a special hole that is made from wax.
- Mosquitoes lay their eggs in water. Mosquito larvae live in water and are quite different to caterpillars that live on land.
- Houseflies lay their eggs in faeces, rotting food and other decaying **matter**.
- Some wasps lay their eggs in the bodies of living caterpillars.
- Some insects hold their eggs inside them and the young hatch inside the mother, for example, aphids and blowflies.
- Silkworm moths spin a cocoon of silk around the pupa which is quite different to the pupa of bees.

For you to try

1 **Investigation: Mosquito life cycle**
   a Collect a few mosquito larvae, and a clear plastic or glass bottle or jar with a lid.
   b Place the mosquito larvae in the bottle with some water and the stem of a plant.
   c Observe the larvae for several days and record your observations. How long does it take for the eggs to hatch?
   d Draw the life cycle of the mosquito.

2 **Investigation: Looking at cocoons**
   a Collect some different cocoons from bushes and walls around your school, some clear plastic or glass bottles or jars, pieces of cloth or net and pieces of string.
   b Place each cocoon in a bottle or jar and tie the cloth or net over the neck of the jar with the string. This is to let air inside but stop the insect from getting out.
   c Place each jar in a cool place and observe the cocoon each day.
   d Draw the adult insect that comes out of each cocoon.
   e Try to identify the name of each insect.

3 What is a life cycle?

4 Explain, using examples, the difference between the three types of life cycle in insects.

Reproduction in vertebrates

- Most **vertebrates** reproduce by sexual reproduction.
- The female produces egg cells in the ovary and sperm cells in the **testes**.
- The sperm cell is smaller than the egg cell. Sperm have a small food supply and move by swimming with their tail.
- The egg cell is larger and has a bigger food supply, but does not move.

There are two types of fertilisation:

1 External fertilisation occurs when the egg and sperm join outside the female's body, for example, fish and frogs.

2 Internal fertilisation occurs when the egg and sperm join inside the female's body, for example, reptiles, birds and mammals.

This results in two types of development:

a External development occurs when the embryo becomes a young vertebrate outside the female, for example, fish, frogs, reptiles and birds.

b Internal development occurs when the embryo becomes a young vertebrate inside the body of the female, for example, most mammals, including humans.

Most birds and mammals care for their young, which is called parental care. Most fish, frogs and reptiles do not care for their young.

Fish

- Eggs develop in the ovaries causing the female to swell.
- The female lays eggs in water. This is called **spawning**.
- The male squirts sperm into the water as he swims over the eggs.

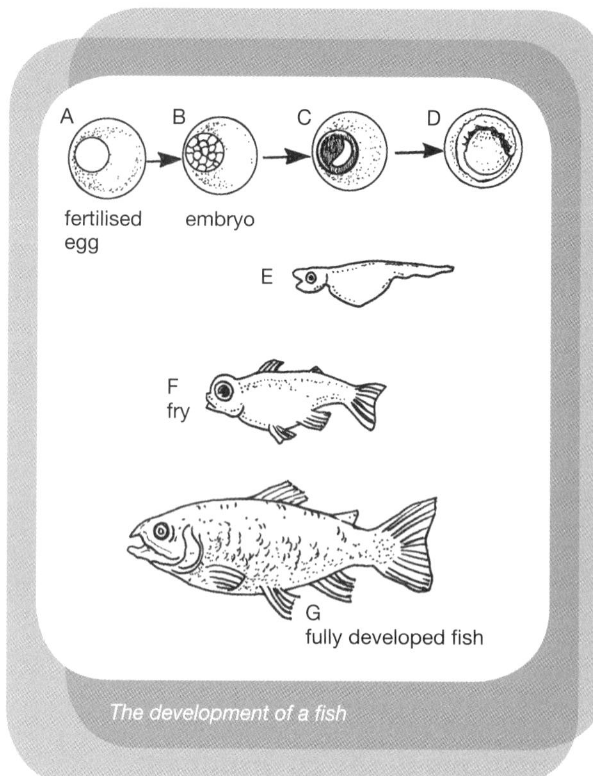

The development of a fish

- Fertilisation is external.
- Fish have external development.
- The egg has a large amount of yolk, which is food for the embryo as it develops. Development may take several weeks.
- Some fish can produce live young, for example, sword tails and guppies.

Amphibians

- The body of the female frog swells as the eggs develop.
- The male calls the female to mate.
- In the breeding season, the thumbs of the male frog become larger which helps him to hold the female.
- The female usually lays eggs in water and the male squirts sperm over them.
- Each egg has a special jelly-like coat which swells in water and is known as **frogspawn**.

This helps to hold the eggs together, protects them from being eaten and is also the first food of the embryo.

- Development of the frog is external.
- The fertilised egg develops into an embryo called a **tadpole**.
- Eyes, mouth and external gills develop and the tadpole begins to swim and feed. The tail grows longer and internal gills develop.

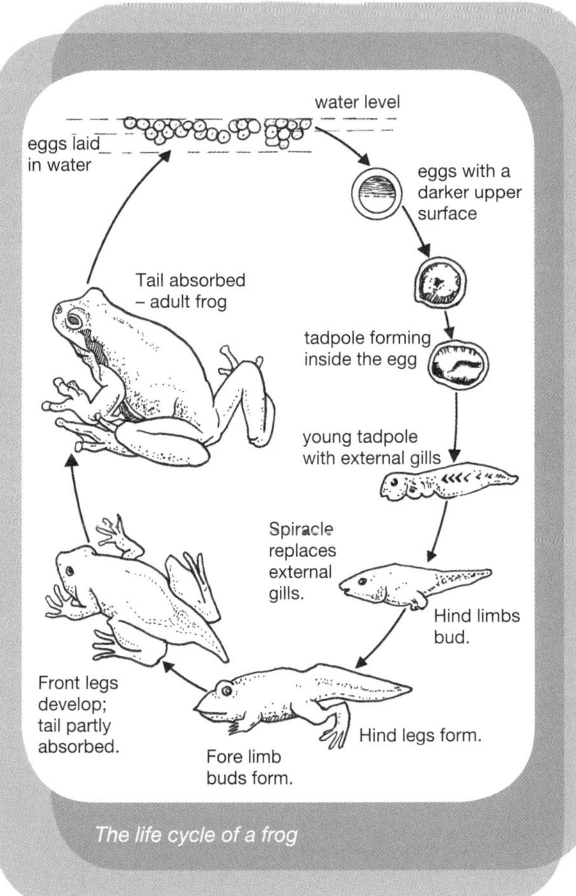

The life cycle of a frog

- After about 100 days many body changes take place which are called metamorphosis.
- The back legs start to develop and then the front legs. The lungs develop and the gills stop working.
- At this time the frog does not eat but the tail is absorbed slowly as food.
- When the tail is gone the young frog leaves the water and moves by crawling, hopping and swimming.
- After about two years the frog can reproduce.

Adaptations of fish and amphibians

Both fish and **amphibians** such as frogs produce many eggs but only a few of these are joined with sperm. Many eggs are eaten by fish before they hatch. After hatching young fish and tadpoles are also eaten by carnivores. Fish and frogs are able to survive as species because of the large number of eggs that are laid. For example, some fish lay several million eggs.

In Papua New Guinea about half of the frogs do not lay eggs in water. Because of the high rainfall and humidity, many frogs are able to lay eggs in moss or leaf litter. Fertilisation and development are still external, but some of these frogs do not go through a tadpole stage.

Reptiles

- Reptiles can live and reproduce away from water and so fertilisation is external.
- Each egg is covered with a tough shell to stop it drying out.
- The egg contains stored food to feed the young reptile.
- The parents do not usually look after the eggs but the warmth of the sun incubates them and development is external.
- Examples of egg-laying reptiles are geckoes, dragon lizards, goannas, pythons, tortoises and turtles.

- In some reptiles development is internal so the female does not lay eggs. The snake develops inside the female and uses the yolk from the egg but does not receive food from the mother.
- Some snakes can give birth to thirty live young at one time. Examples are death adders and sea snakes.
- Most reptiles do not care for the young and they must look after themselves.

Birds

- Female birds have only one ovary which reduces their weight for flying.
- Fertilisation in birds is internal and occurs when the male treads on the back of the female.
- Birds lay eggs with a **hard** shell which protects the egg and prevents the loss of water, but still allows air to pass through the shell to the embryo. The oxygen in the air is needed for respiration.
- After the egg has been laid it needs to be kept warm or incubated so the egg will develop. The female usually sits on the eggs although brush turkeys use the heat from rotting plants to keep the eggs warm and in some places the megapode uses the natural heat from rocks deep underground.
- The yolk of the egg provides the baby bird with food.
- Hatching begins when the baby bird picks a hole in the shell.
- The number of eggs laid by birds depends on the parental care and amount of development before hatching.
- Birds such as hawks produce few eggs, the young hatch at an early stage and need a lot of care, including regular feeding.
- Birds such as ducks produce more eggs, the young are more developed when they hatch and soon leave the nest to take care of themselves.

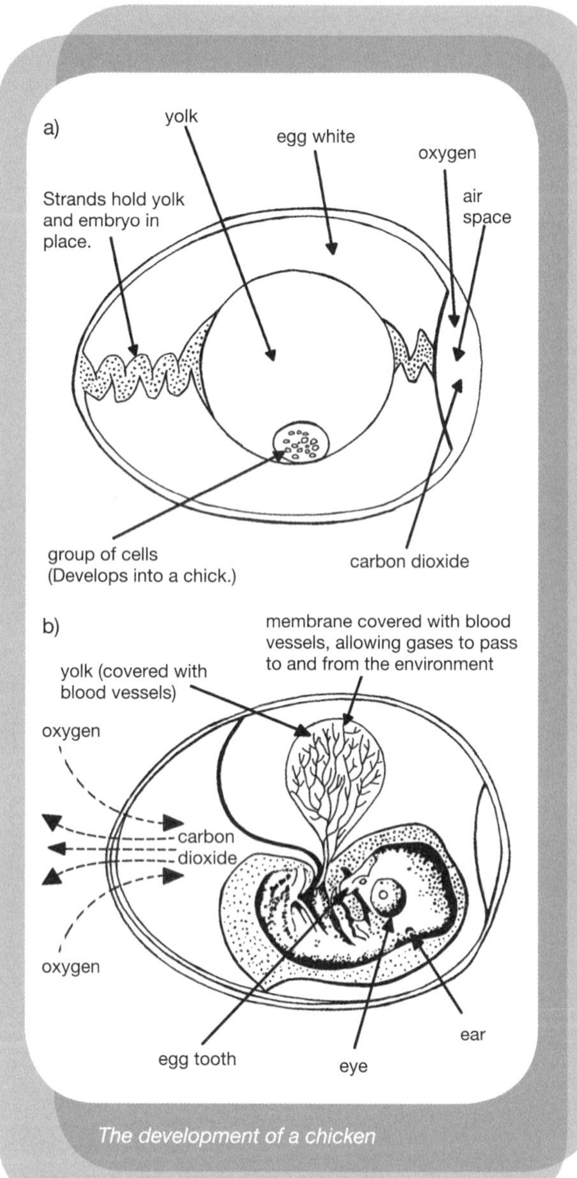

The development of a chicken

- Birds can lay eggs that are not fertilised. For example, the chicken eggs that are sold in stores for people to eat are not fertilised.

Mammals

- In most mammals fertilisation and development are internal.
- Between one and ten eggs are produced at a time. Fewer eggs are needed as more young survive.

- Male mammals have a penis that leaves the sperm near the eggs and so the chance of fertilisation is high.
- The young of most mammals are well developed and quite strong when born, parents care for their young and feed them on mother's milk. This increases their chances of survival.
- The period of development varies with the species as shown in the table right:

The period of development of some mammals

Name	Average time (days)
Black rat	21
European rabbit	31
Cat	63
Dog	63
Pig	114
Sheep	151
Human	278
Horse	336
Indian elephant	624

Groups of mammals 

There are three main groups of mammals:

1 Monotremes such as the platypus and echidna which lay eggs that have shells.

2 Marsupials such as the wallaby, kangaroo and bandicoot, which have young that are born prematurely. The female has a pouch and the embryo slowly moves to the pouch where it feeds on milk and keeps warm as it develops.

3 Placentals which have young that complete their development inside the female. The embryo is joined to the mother by the **placenta** which provides food and oxygen and also gets rid of waste. Examples of placentals are cats, dogs, pigs and humans.

Placental mammals

- Eggs are formed in the ovary of the female and are fertilised in the egg tube or **uterus**.

Development and birth in the kangaroo: a) Female about to give birth b) Embryo moving in the pouch c) Embryo attached to a nipple in the pouch

- Humans have only one uterus but some placentals have two uteri.
- The embryo is attached to the uterus by the placenta which passes food and oxygen from the mother to the embryo and carries waste from the embryo to the mother. The **umbilical cord** connects the embryo to the placenta.
- The young embryo is well protected in the uterus until it is fully developed and ready to be born.
- During birth the baby passes out through the **vagina** and the umbilical cord is cut.
- Humans usually have only one baby at a time.

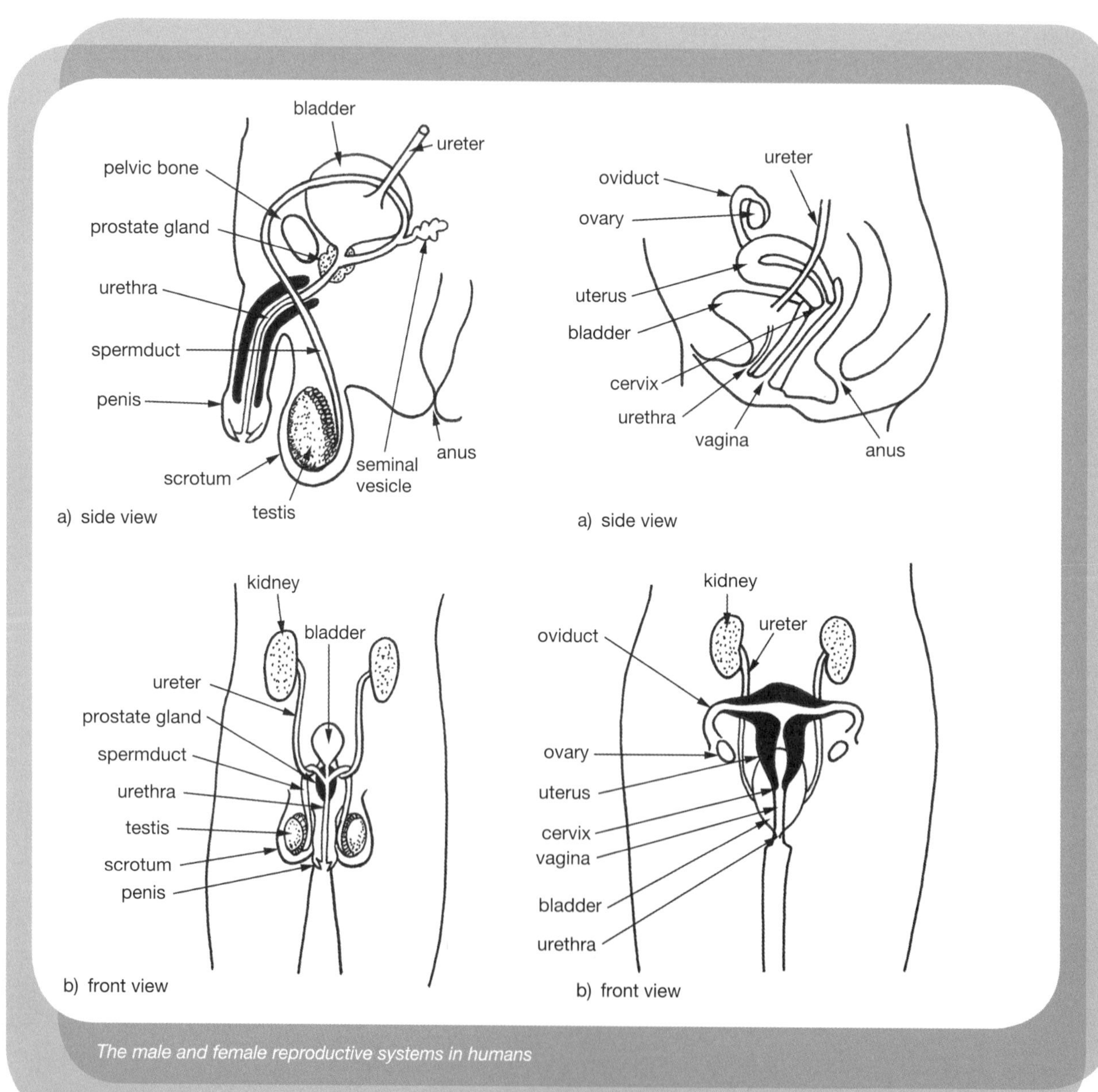

The male and female reproductive systems in humans

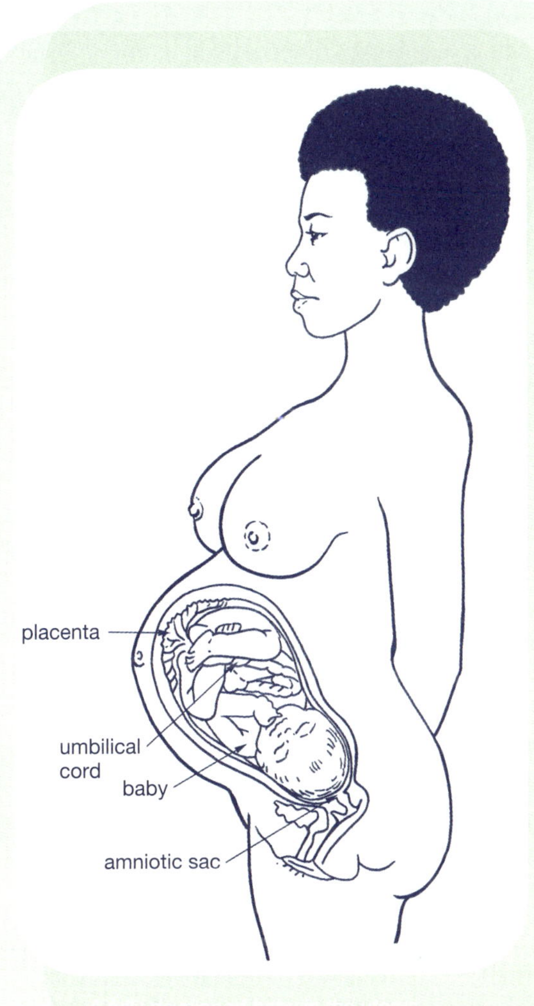

A fully developed baby in the uterus

For you to try

1 **Investigation: Life cycle of the frog**

   a Collect some fresh frogspawn and a container such as a large tin, dish or bowl.

   b Put the frogspawn in the container and place it in a cool place where it will not be disturbed.

   c Make sure that there is enough water to stop the frogspawn drying out and add water from time to time if needed.

   d Observe the frogspawn each day and record any changes in a diary or log book.

   e How long does it take for the tadpoles to change into young frogs?

   f Draw the life cycle of your frog on a sheet of paper showing the time taken for the different stages in the development that you observed.

   g Put the frogs back where you found them.

2 **Investigation: Life cycle of the gecko**

   a Find a clear plastic bottle or container.

   b Make some small holes in the plastic bottle to allow air to go inside.

   c Look in cracks and small holes near doors and behind walls to find some gecko eggs.

   d Carefully put the eggs in the plastic bottle without breaking them.

   e Observe the eggs each day and record any changes in a diary or log book.

   f How long does it take for the eggs to hatch into adults?

   g Draw the life cycle of your gecko on a sheet of paper showing the time taken for the development that you observed.

3 Frogs use external fertilisation and lizards use internal fertilisation. Explain the difference. 

4 Fish have external development and pigs internal development. What does this mean?

5 Describe the way in which the development of the young differs in the three main groups of mammals. 

(continued over)

6 Look at the table with the heading **Summary of reproduction in vertebrates** (below). Describe the relationship between the number of young produced, the amount of development before birth and parental care afterwards.

Projects

A

1 Make a small pond in your school grounds in a place that is suitable and safe. Collect some water plants and small fish and put them in the pond. A variety of plants and animals is best, but after a while other animals like frogs and insects like dragonflies will come to the pond by themselves. Observe your pond every day and keep a record of your observations in a diary or log. Follow the life cycle of different plants and animals. If you observe your pond for many months you should be able to follow the seasons of different plants and animals and see the relationships between them. If you are not able to make a pond of your own you can still follow the life cycles of plants and animals by looking at a small pond, creek or river nearby.

The table below summarises reproduction in the five classes of vertebrates.

Summary of reproduction in vertebrates

Class	Size of eggs	Number of eggs	Shell	Fertilisation	Where are eggs laid?	Parental care	Food of new born
Fish	small	thousands	none	usually external	water	none or very little	yolk sac
Amphibian	small	hundreds or tens	none	external	usually water; sometimes land	none or very little	yolk sac
Reptile	large	few to tens	leathery shell	internal	land	almost none	yolk sac
Bird	very large	1–15	hard shell	internal	land	for a short time	yolk sac; then seeds and insects
Mammal	microscopic, except platypus	1–20	none, except platypus and spiny anteater	internal	developed in mother	for a long time	milk

Ecology, relationships and interactions

Using materials A

People have always used materials to help them in their way of life. In the past people wore clothing made from plants and animal skins. They made tools from wood and stone or the bones of animals. They cut down timber to build houses and canoes and to get firewood. They also learned to grow crops and look after animals for food. They made pots from clay so that they could cook their food. As time passed they found more materials and learned to make them into useful objects. For example, they learned to use **metals** to make tools and cooking pots. They also worked out how to use fibres from plants and animals to make bilums and weave fabrics to make clothes and blankets.

Natural and synthetic materials

Today people have a much larger number of materials to use for their various needs than in the past. We have learned new ways of using materials but we have also found ways of making new materials. The **natural materials** that we use come from various sources:

- Plants—cotton, linen, jute, wood and rubber
- Animals—wool, leather, silk, fur and bone
- Earth—stone, clay, copper, aluminium and gold

The substances that are used to produce other materials are called **raw materials**. Materials that people have made or manufactured such as flour, cement, cordial and paint are called **processed materials**. Processed materials that are made from chemicals or artificial substances are called **synthetic materials**. For example, many plastics, nylon and polystyrene are made from chemicals.

For you to try

1 Investigation: Sorting materials

- **a** Collect various examples of materials such as flour, sugar, milk, wood, cotton, wool, a nail, a stone, clay, a marble, a pen top etc.
- **b** Identify the main material that each is made of.
- **c** Decide whether the material is natural or processed.
- **d** Where does each material come from?

(continued over)

e Describe each material by using things such as size, shape, colour, feel, smell and sound when tapped.

f Record your results in a table like the one below:

Example	Main material	Natural/Processed	Description
Flour			
etc.			

Properties of materials

Different materials have different properties that make them suitable for some uses and not for others. The properties of materials include the way they look, feel and behave.

Strength

Materials that can keep their shape when they are twisted, pulled or crushed are described as strong. We can also increase the strength of materials if they are made into particular shapes. For example, the triangle is a very strong shape and the frame that holds the roof of a building makes use of this. Corrugated iron is much stronger than flat iron because of the shape. If you look at a sheet of corrugated iron from the end you can see that the shape of the corrugations is more like a row of triangles.

There are other shapes that are also very strong. For example, a tube is stronger and lighter than a bar of the same size. Steel beams or joists in the shape of an 'I' or 'H' are often used in buildings with several levels to support the floors above. These beams are known as I-beams and are very strong because of their shape.

Hardness

Hardness is the property of materials that makes them difficult to scratch and stops them wearing away. Harder materials can be used to cut softer materials. Many tools are made from steel because it is hard.

Flexibility

A material that is **flexible** may be bent easily and a rigid material keeps its shape or breaks when you try to bend it. For example, a hose pipe is flexible and concrete is rigid.

Conductors of heat and electricity

Materials that let heat or electricity pass through them very easily are good **conductors** and those that do not let heat or electricity pass through them easily are called **insulators**. Metals are usually good conductors. Wood and plastics are examples of insulators.

Elasticity

Materials that are elastic can be stretched and then return to their original shape. Rubber is the most well-known elastic material.

Corrugated iron roofing and an I-beam

Waterproof

Materials that are **porous** will allow water to pass through them and materials that are waterproof will not allow water to pass through them. For example, the paper used to make a tea bag allows the hot water to pass through to the tea leaf inside so that we can then taste the tea in the cup. A raincoat or umbrella is usually made from plastic material or rubber to stop water passing through it.

An umbrella is waterproof but a tea bag is porous.

Flammable

Some materials burn when heated and those that burn easily are called **flammable**. Different materials behave in different ways when they burn. For example, cotton burns quickly, wool burns slowly and nylon melts while hot sticky pieces fall off. Some materials burn with little or no smoke and others produce a large amount of poisonous smoke. Many synthetic materials like plastics give off poisonous smoke and should not be burned.

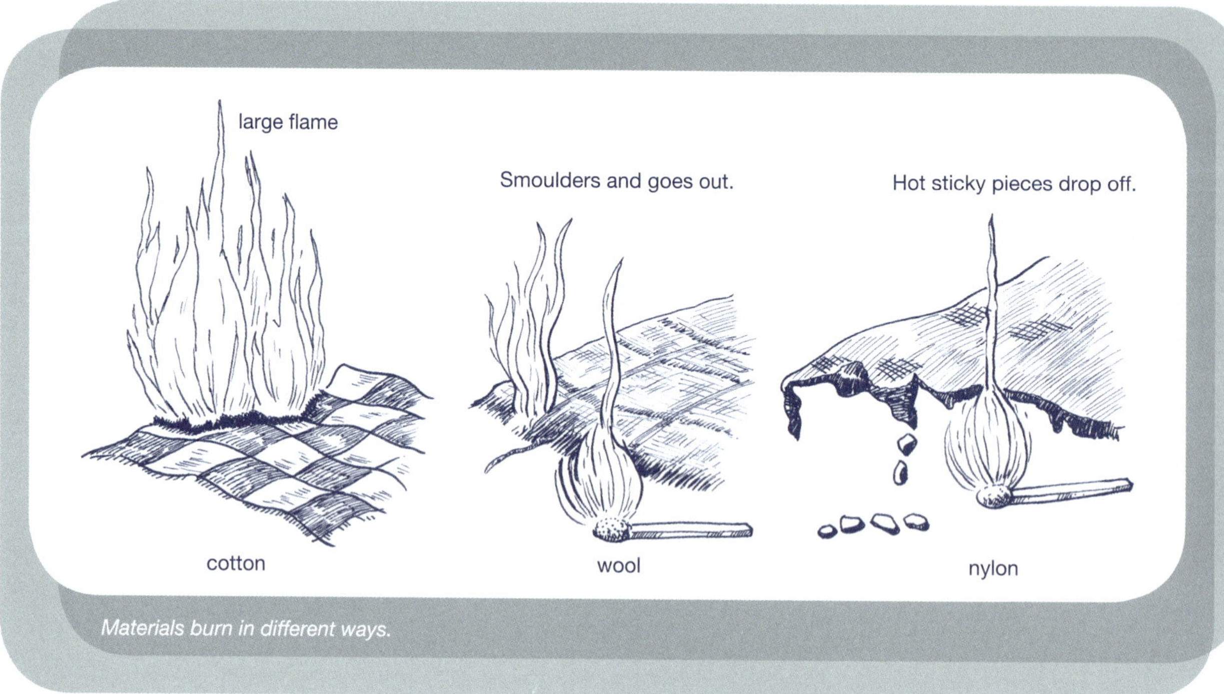

Materials burn in different ways.

Durability

Materials that are **durable** do not wear out easily and so can last for a long time. Things that are made from iron or steel can rust or corrode. Metal **corrosion** can be prevented by covering the surface with a coating that keeps out air and water. For example, corrugated iron is covered with a thin coating of zinc which protects the iron. Metals can also be protected by painting or covering them with a coat of oil or grease.

Some materials, such as cotton, wool and wood can be attacked by pests. For example, moths can eat small holes in cotton and wool, and timber can become rotten because of a fungus. Timber can also be eaten by insects. People sometimes put moth balls, which are made from naphthalene, in a suitcase or clothes cupboard to stop moths making holes in their clothes. Timber can be treated with a chemical solution to help stop it rotting or being eaten by insects. This is known as **treated timber**.

For you to try

1. **Investigation: Using the properties of materials** SR A
   a. Work in groups of four or five.
   b. Look around your classroom and make a list of ten materials and say what each is used for and the properties of the material that are being used. Use the list of properties in this section to help you.
   c. Go outside and make another list of materials that you find. How many

different materials does the second list contain? Why is the second list different from the first list?

d Show your results in a table like the one below. One example has been completed for you.

Example	Main material	What is it used for?	Properties
Roof	iron with a coating of zinc	keeping out the rain and the sun	waterproof, strong, durable

Renewable and non-renewable materials

All materials can be classified into two main groups:

Renewable materials are those that can be replaced because they come from things we grow. For example, natural materials like wood, cotton, wool, leather and rubber are renewable.

Non-renewable materials are those that cannot be replaced because they do not come from the things we grow. For example, metals, stone, clay, and oil.

As more people use them, some non-renewable materials will become more difficult to obtain and eventually there will be no more. This will also affect the way that we use synthetic materials because they may be made from a mixture of renewable and non-renewable materials. For example, clothes are often made from material called polyester cotton, which is a mixture of 65% cotton and 35% polyester. The cotton comes from a plant which is renewable, and the polyester is made from chemicals that are not renewable.

Biodegradable and non-biodegradable materials

Whenever we use materials we also produce waste. The waste substances may be produced when we are using the material or when we have finished using the material. Waste can be a problem and the way that we deal with waste depends on the materials from which it was made. Many natural materials will decay and slowly disappear. Materials that can be broken down by small living things like bacteria and fungi are said to be **biodegradable**. Wood, paper and cardboard cartons are all biodegradable. Materials that are not broken down by small living things like bacteria and fungi are said to be **non-biodegradable**. For example, tins, bottles, plastic, batteries and rubber tyres will all last for many years, even after we have stopped using them.

For you to try

1 Make a list of the things that you use in your everyday life and classify them in a table as biodegradable or non-biodegradable.

2 Copy and complete the table below to show examples of the waste substances that are produced when you use natural or processed materials, and waste that is left after you have finished using the materials. Two examples have been completed for you.

Material being used	Natural or processed	Waste produced during use	Waste left at the end
Firewood	natural	smoke	ash
Battery or dry cell	processed	none	metal case with chemicals inside
etc			

Getting rid of waste materials and staying healthy

The way that we get rid of waste depends on whether the material is biodegradable or non-biodegradable. Biodegradable materials will rot or decay and the smell of decay attracts flies which can spread germs to people or food. When things decay we can often see their colour change and we might also see the shape of the material change. The speed of decay will depend on the material and on the conditions. Some materials will decay very quickly while others will be much slower. Warm and damp conditions usually help the bacteria and fungi to break down the material. Some of these bacteria and fungi may make people sick and so the way that we get rid of waste materials can help us to stay healthy.

Biodegradable rubbish such as leaves, grass cuttings, vegetable peelings and food scraps can all be put on a compost heap and allowed to rot or decay. After some months the compost can be spread on the garden to help improve the fertility of the soil.

Paper and cardboard can also be added to a compost heap or burned in a special place. However, fires should not be lit at the base of trees, as this will kill the tree.

Non-biodegradable materials and objects made from them such as tins, bottles and plastic should be buried in a hole in the ground and covered when the hole is nearly full. Tins and bottles should not be left lying around because they can cut peoples' feet and collect water which allows mosquitoes to breed. Plastics should not be burned because many of them give off poisonous gas and smoke which makes it hard to breathe and can make a person sick. In some countries biodegradable plastic bags are now being made.

People who live in towns or on government out-stations may have their rubbish collected by the local council; it is then taken to a special dump.

Different ways of getting rid of waste substances

For you to try

1 **Investigation: Biodegradable and non-biodegradable** SR

- a Collect various materials such as fruit, vegetables, flour, sugar, milk, wood, cotton, wool, a nail, a stone, clay, a marble, a pen top, a plastic bag etc.
- b Plan and conduct some tests to find out if the materials are biodegradable or non-biodegradable.
- c Think about the senses that you will use when you make your observations.
- d Remember that it may take time before you get results.
- e Record your results in a table and make some conclusions.
- f Write up a scientific report to say what you did and what you found out.
- g As a long-term investigation leave a plastic bag outside in the sun for many months in a place where it will not be disturbed or blown away and observe what happens.

2 With your teacher, invite a guest speaker to give a talk about the effects of biodegradable and non-biodegradable materials in the community. PD

Conserving materials

When we **conserve** materials we use them very carefully so that they will last as long as possible. For example, selective logging is a way of conserving resources. It is also important that we do not waste materials. For example good quality timber should be used to make furniture or build a house, not to make matchsticks or paper.

Recycling

Some materials can be used again or recycled instead of being thrown away. **Recycling** is another way of conserving materials. People in Papua New Guinea recycle plastic containers to carry water or kerosene and people who sell fruit and vegetables in the market also have recycled plastic bags and rice bags for sale. Secondhand clothes shops are very popular in many countries and this is a good way to recycle clothing. Glass bottles and plastic bottles can also be recycled and provide a way for people to earn some money.

Metals that are mined from the ground can also be used again. This conserves the metal ore but also saves large amounts of **energy**. For example, people often collect old aluminium cans for recycling. The cans are melted down into large blocks which can then be used to make new aluminium products. This reuses the aluminium, provides an income for the people who collect the cans, and helps to keep the environment clean and tidy.

Scrap metal, which is mainly iron and steel, is also collected in Papua New Guinea and shipped overseas where it is melted down and reused. In some countries special rubbish trucks come round to peoples' houses to collect paper, glass, aluminium, and some plastic for recycling.

In Port Moresby aluminium is recycled by being melted down in a furnace and then reused.

For you to try

1 Investigation: Recycling paper

a Collect a piece of paper or carton (about one square metre), an old dried milk tin and a piece of flywire (about 15 cm by 10 cm), a plastic bag, and a glass jar. Each group can use a different kind of paper or carton, such as newspaper, butcher paper, brown paper, cardboard etc.

b Cut the paper into about twelve pieces and place them in about one litre of water in the milk tin. Stir the paper until it breaks up and the fibres are evenly spread through the water. (This is a slow process and another way of doing it is to soak the paper or carton overnight.)

c Dip the piece of flywire into the water containing the fibres, carefully lift it out and shake it gently to make the fibres settle and drain some of the water.

d Place the flywire onto several sheets of newspaper and then cover it with a piece of plastic bag.

e Roll the flywire with the glass jar to squeeze out most of the water.

f Carefully remove the plastic and lift the flywire from the newspaper making sure that the fibres stay on the flywire.

g Allow your sheet of fibres (paper) to dry.

h Compare the papers made by different groups. How are they different?

2 Plan and carry out a survey to collect information about the way that people use materials in the community and the way that they deal with waste materials. Analyse the information and give some feedback to the community about the wise use of non-biodegradable materials.

3 Make a poster for your school to encourage students to 'reuse, reduce and recyle'.

4 With your teacher, invite a guest speaker to talk about what can be done to reduce the impact of human activities such as mining, fishing, logging, building roads and factories.

PD

Chemical pesticides

A pesticide is a chemical that is used to kill a pest, weed or disease:

- insecticides are used to kill insects, for example, DDT, malathion, pyrethrin
- herbicides are used to kill weeds, for example, paraquat
- fungicides are used to kill fungi, for example, PCP used for treated timber (also kills insects).

People use pesticides so that crops and animals can survive better and to stop timber rotting or being eaten by insects. Unfortunately, the pest is not the only living thing that is affected by the pesticide. For example, pesticides can be passed along food chains, and insecticides can kill useful insects as well as insects that are pests.

Scientists in the United States of America have shown that the insecticide DDT could be detected in all the organisms in a food web of a river mouth. They found that the amount of DDT increased along the food chains, as shown in the diagram below. The reason why more DDT is found in the clam than the plankton is because the clam consumes a great deal of plankton. Nearly all of the DDT is stored in the clam's body. The gull eats many clams and stores most of the DDT in its body.

Food chain showing the way that DDT increases in animals along the food chain

The DDT has had serious effects on the birds. It makes birds lay eggs with very thin shells which break when the adult sits down to incubate them. So very few young are produced and the population decreases as adults die and are not replaced by the young.

The effects of DDT have been so serious in some countries that many birds of prey such as eagles, hawks and ospreys have become rare.

DDT has also been found in human milk and body organs but we do not know how dangerous this substance is to humans. The long-term effects are not completely understood.

Another problem with insecticides is that insect populations are becoming resistant to the chemicals. Some of the insects are naturally resistant to the chemicals and will survive. Other insects move into the area after spraying, some will breed with the resistant ones and some may inherit the resistance. When they are sprayed again more insects will be resistant and will survive. With continued spraying, all of the population may become resistant.

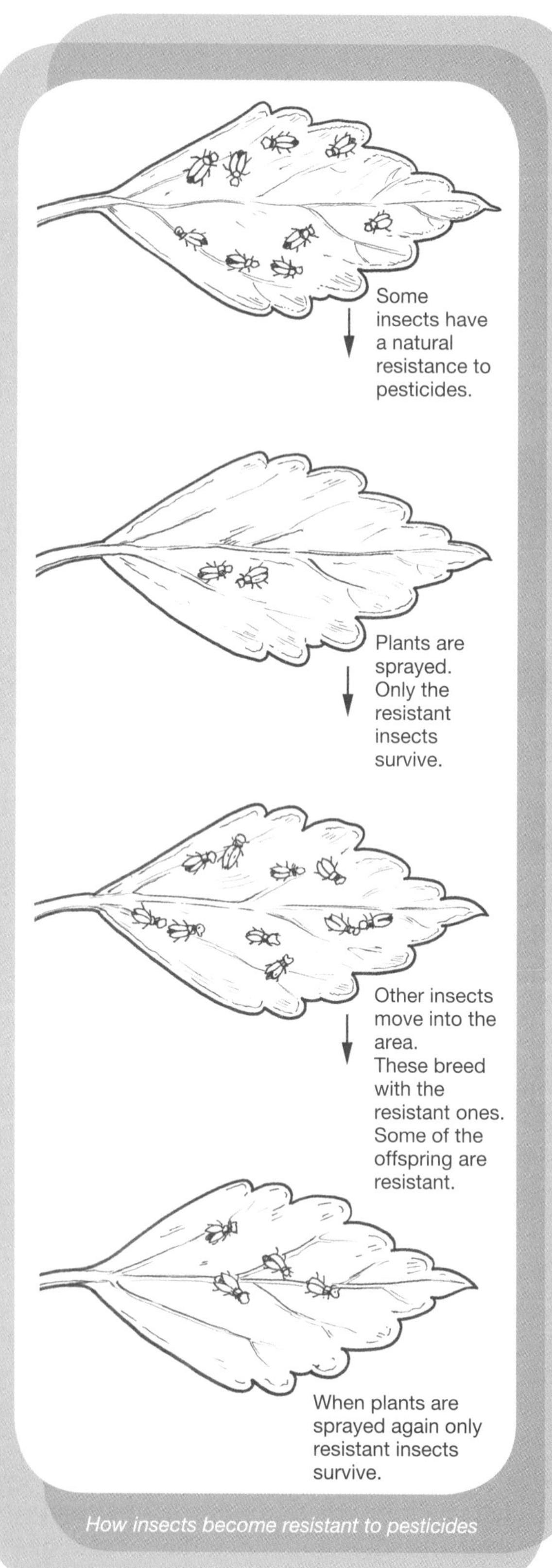

How insects become resistant to pesticides

Chemical pollution

Dangerous chemical wastes from factories and mining are known to have an effect on people. One of the most well-known cases comes from Minamata Bay in Japan where people began to get sick in the 1950s. The problem was found to be caused by mercury in the waste water that flowed into the bay from a chemical factory. Poisonous mercury compounds had been concentrated in the food chain in the bay, like the example of DDT mentioned previously. People eating fish and shellfish had absorbed large amounts of these mercury compounds. Cats eating the food scraps also got sick and died. By 2001, 2265 victims had been identified, 1784 of whom have died, and the company had paid compensation to over 10 000 people.

Damage caused by waste materials from the Ok Tedi mine

In Papua New Guinea people are concerned about the chemical waste in the **tailings** that mining companies release into the rivers.

Chemical pollution can be avoided by:

- having proper sewage treatment to make human waste safe before it is allowed to run into rivers or the sea
- using tailings dams to collect and hold back chemical waste so that it is not allowed to run into rivers while it is still poisonous
- using safer gases in refrigerators—safer gases will not destroy the ozone layer and should replace the chemicals called chlorofluorocarbons or CFCs
- using safer gases in spray cans—safer gases such as hydrocarbons will not destroy the ozone layer
- passing laws to control the type and amount of chemicals released into the environment.

For you to try

1 Explain how poisonous waste materials carried in rivers can affect people who eat fish and shellfish from the sea.

Projects

L

Start a recycling project in your school or community. Collect useful things that other people are throwing away and reuse them in the school or sell them to raise money for the school.

Misima cyanide spill

In August 2004, there was a report of a cyanide spill into the ocean on Misima Island that resulted in a number of fish being killed. The spill happened when cyanide tanks were being cleaned out as the Misima mine was being closed down. The mining company said that only 35 fish were killed but local people reported that the spill had a much greater effect on the environment. A whale was reported washed up on shore near the tailings outfall. There were also reports from the outer islands of turtles, sharks, yellow fin tuna, flying fish and sardines being killed.

People use the sea water for cooking and washing but became scared to use it and afraid of handling fish. People were also worried about what the mine had done to the oceans through the tailings that had been dumped in the seas during the 15-year life of the mine.

Summary questions

1 Look at the following statements about reproduction and answer true or false:

Statement	T or F?
Reproduction is one of the characteristics of living things.	
All living things eventually die so reproduction is needed for the species to survive.	
Every individual plant or animal needs to reproduce in order to survive and complete the life cycle.	
Different species reproduce in different ways, depending on the environment in which they live.	

2 Which of the following is **not** an example asexual reproduction?

- **A** simple or binary fission
- **B** budding
- **C** producing spores
- **D** pollination
- **E** vegetative reproduction

3 Which of the following can both take place during sexual reproduction?

- **A** adaptation and regeneration
- **B** pasteurisation and pollination
- **C** fertilisation and pollination
- **D** fertilisation and spontaneous generation

4 Choose the best matches in the following pairs:

A	The joining of a male sex cell with a female sex cell	**I**	External development
B	Two sex cells join together outside the female's body	**II**	Internal fertilisation
C	Two sex cells join together inside the female's body	**III**	Internal development
D	The young animal grows inside the mother after fertilisation	**IV**	Pollination
E	The young animal grows outside the mother after fertilisation	**V**	Fertilisation
F	The movement of male sex cells in plants	**VI**	External fertilisation

5 Look at the following statements about seed dispersal and answer true or false:

Statement	T or F?
Plants usually scatter their seeds close to the parent for protection.	
Dispersal of seeds reduces the competition for water, sunlight and minerals.	
Seeds that are eaten by animals must be digested before they pass out with the droppings and start to grow.	
Seeds that are carried on the fur of animals usually have hooks or burrs.	
Seeds that are dispersed by water always sink.	
Mechanical dispersal is more likely to happen on a hot, dry day than on a cool, wet day.	
Seeds that are dispersed by wind usually have wings or parachutes.	

6 Which of the following best describes the order of the stages in the life cycle of insects that have complete metamorphosis?

A adult ➔ egg ➔ larva ➔ pupa

B adult ➔ larva ➔ pupa ➔ egg

C egg ➔ larva ➔ pupa ➔ adult

D egg ➔ pupa ➔ larva ➔ adult

7 Which of the following statements about mammals are true compared to other groups of vertebrates?

I produce fewer eggs

II usually more developed before they are born

III show more parental care

A I and II

B II and III

C I and III

D I, II and III

8 Look at the following statements about the use of materials and answer true or false:

Statement	T or F?
All natural materials are renewable materials.	
Materials that are renewable come from things that grow.	
Non-renewable materials usually come from non-living things.	
Things that are biodegradable usually last longer than non-biodegradable object.	
Things that are biodegradable are usually made from renewable materials.	
All synthetic materials are processed materials.	
All processed materials are synthetic materials.	

9 The passage opposite is a summary of the main ideas of this chapter. Copy and complete the passage in your book. Using the words in the list, find the words that are missing. You can use each word only once.

biodegradable, cycle, environment, external, fertilisation, individual, internal, non-biodegradable, parent, reproduce, reproduction, rot, sexual, survive, wisely

All living things pass through a life ____________ which consists of different stages.

At some stage in the life cycle living things are ready to ____________. Reproduction is necessary for the species to ____________, but an individual plant or animal does not need to reproduce in order to complete its own life cycle. Reproduction can be asexual or ____________. In asexual reproduction the new individual is produced from one ____________ only. There are different types of asexual ____________. In sexual reproduction the new ____________ is produced from two parents. During sexual reproduction the male sex cell and the female sex cell join together during the process of ____________. Fertilisation that occurs outside the body of the female is called ____________ fertilisation and fertilisation that occurs inside the body of the female is called ____________ fertilisation. Different living things reproduce in different ways depending on the ____________ in which they live.

All living things will eventually die then ____________ or decay. Materials that will rot or decay naturally are called ____________ materials. Materials that will not rot or decay naturally are called ____________ materials. We need to use all materials ____________, and understand how to get rid of non-biodegradable materials that are no longer useful.

3

Science in the home

Chapter summary

In this chapter you will have an opportunity to:

- find out about the properties of different materials and make models to show the differences in properties
- find out about substances called acids and bases and how they can be used
- find out more about the ways that we use electricity and heat
- find out more about simple machines and the ways that they can be used in homes and the community to make life easier.

Syllabus references

Strand:	Science in the home	
Sub-strand:	Learning about substances Using energy in the home	
Outcomes:	**8.3.1**	Explain the structure of matter in terms of the particles from which it is made
	8.3.2	Identify and collect basic and acidic substances found in nature and use these data to elaborate on how they can be used to benefit the community
	8.3.3	Apply their knowledge about energy to investigate electrical and heat energy in the home
	8.3.4	Apply their knowledge about force to investigate simple machines
	8.3.5	Conduct investigations on simple machines and use problem-solving skills to establish the efficiency of the machine as a tool to do work

Key facts

- The materials that we use every day have different physical properties such as hardness, strength, density.
- When we use materials we choose those that have the most suitable properties. For example, some things will float and some things will sink.
- Some substances have different chemical properties. For example, some things are acids and some things are bases.
- Acidic and basic substances can be found in nature and are used by people in different ways.
- Acids and bases can also be made and used in everyday life.
- When we use energy in the home we often use heat energy and electrical energy.
- Heat is a form of energy that can make substances hot and can be controlled.
- Temperature is a measure of how hot a substance is.
- Electricity has three main uses or effects—heating, lighting and magnetism.
- Electricity must flow around a path called a circuit which we can draw as a diagram.
- Simple machines can help people to do work because there is a mechanical advantage.
- We can calculate the amount of work that we do and the mechanical advantage of different machines.

Learning about substances

Properties of materials

The characteristics of a material and how it behaves are called its **physical properties**. People use different materials for different purposes. The choice of material is usually based on the physical properties of that material. For example, copper is a soft metal that bends easily and does not rust so we can use sheets of copper on the bottom of a wooden boat to protect the boat, and to make pipes to carry water. We can also use copper to carry electricity because it can easily be stretched into wire and is a good conductor.

Window louvres can be made out of wood, metal or glass. We use glass louvres because they let light through. Plain glass is **transparent** and we can see clearly through it. Some glass has a pattern on it and is called **translucent**, opaque or obscure. Obscure glass will let light through but not let people see through, which is a useful property. However, not all properties are useful. For example, glass is also **brittle** or can break easily so we have to be careful and may have to replace broken louvres. Metal and wooden louvres are stronger and do not break so easily but we cannot see through them.

Some of the words used to describe the physical properties of different materials are shown below:

Physical property	Meaning	Examples
Strong	resists the effects of forces	steel, concrete, timber
Hard	not easily scratched or worn away	steel, some rocks, glass
Brittle	hard, but can break easily	glass, fibro, dried clay
Malleable	bends easily and can be hammered into shape	copper, aluminium
Ductile	can be pulled out into wires	copper, aluminium
Lustre	shiny or lustrous	most metals
Transparent	'see-through' or lets light through very easily	clear glass, perspex, clear plastic
Translucent	lets some light through, but scatters it	obscure glass, white polythene
Conductor (heat)	lets heat pass through easily	copper, aluminium, iron, steel
Conductor (electricity)	lets electricity pass through easily	copper, aluminium, iron, steel
Insulator (heat)	stops heat passing through	expanded polystyrene
Insulator (electricity)	stops electricity passing through	wood, pvc, glass
Dense	a lot of matter in a given space—dense objects sink in water	metals are dense, cork and polystyrene are not dense

Examples of objects made from different kinds of materials

For you to try

1 Investigation: Properties of matter SR A

- **a** Collect a selection of objects made from different materials to test: a rubber band, a piece of chalk, a stone, a glass marble, an aluminium can, a nail, a piece of corrugated iron, a piece of plastic, pieces of wire, a wooden stick, a block of wood, a piece of cloth, a candle etc.
- **b** Carry out some tests to find out about the properties of the materials. Use the table above to help you.
- **c** Write up a scientific report to say what you did and what you found out. Include your results in a table, put the list of materials in the first column and then have a separate column for each property.
- **d** What conclusions can you make about different materials?

States of matter

All materials are made of matter and can be divided into three groups called solids, liquids and gases. These are known as the **states** of matter. The state of matter depends on the temperature and the pressure. Some materials can change their state depending on the temperature and pressure. For example, liquid water will change to solid ice at a temperature of 0°C. Water will also change to steam at a temperature of 100°C. However, when we heat water on a high mountain it will boil at less than 100°C because the pressure of the air is less.

One of the physical properties of the air in the **atmosphere** is called **atmospheric pressure**. At sea level, the particles of air push down with a force equal to one kilogram pushing down on each square centimetre. In other words, if we imagine a column of air that measures one centimetre by one centimetre but which is 400 km tall, the air in this column will weigh one kilogram. We do not feel the **mass** of the atmosphere because it acts equally in all directions. Our bodies are also made to withstand normal pressures and push back with an equal pressure.

As you move further away from the Earth's surface there are fewer particles in the same volume of atmosphere and we say that the air is thinner. Climbers and people who live in high mountains know that it is more difficult to breathe there because the air is thinner.

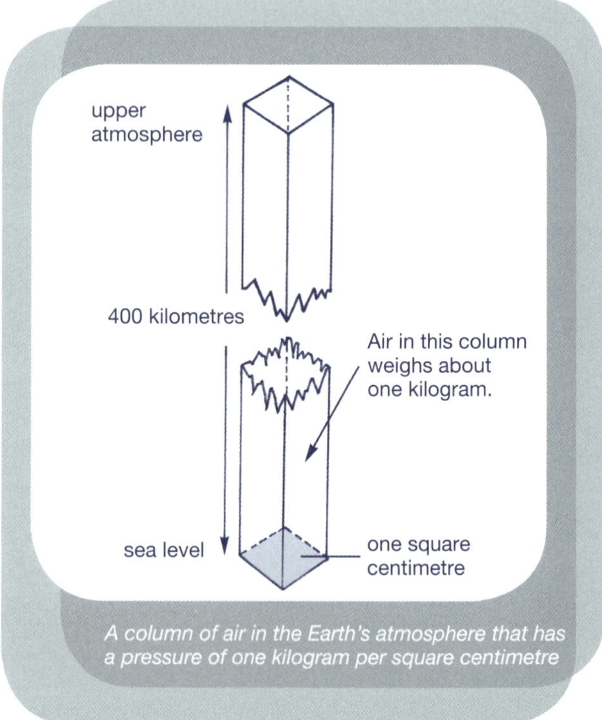

A column of air in the Earth's atmosphere that has a pressure of one kilogram per square centimetre

For you to try

1 **Investigation: Showing air pressure (1)**
   - a Collect a glass and a piece of smooth cardboard.
   - b Fill the glass completely with water and place a piece of stiff cardboard over the top.
   - c Hold the cardboard in place and turn the glass upside down.
   - d Take your hand away from the cardboard and observe what happens.
   - e The pressure of the air pushing against the cardboard, in this case acting upwards, is greater than the weight of the water which acts downwards.
   - f As a result the cardboard stays in place and the water does not fall out of the glass.

2 **Investigation: Showing air pressure (2)**
   - a Collect a glass or plastic drink bottle and a bucket of water.

b Fill the bucket with water and hold the bottle under the water until it is full of water.

c Hold the bottom of the bottle and carefully lift it out of the water but always keep the neck of the bottle under the water in the bucket.

d Observe what happens to the water in the bottle.

e Write up a scientific report to say what you did and what you found out.

Density

Some materials sink and some things float. For example, stones always sink but corks will always float. This is not because they are heavy or light, because even stones that are light will sink. Materials will float or sink because they are heavy or light for their size. The heaviness of an object for its size is called its **density**. In other words density is a measure of the amount of material in a given space.

Many kinds of timber will float but some will sink. Timber companies sometimes float logs in the sea before they are loaded on to a ship with a crane, especially where there is no wharf. The logs that sink must be tied to the ones that float or else they will be lost.

For you to try

1 Investigation: Floating and sinking  

a Collect a number of different materials such as small pieces of different kinds of timber especially hardwoods, different metal objects, different kinds of stone including a piece of pumice, some cork and a bucket of water.

b Test each object to see if it will float in water.

c Write up a scientific report to say what you did and what you found out. Show your results in a table.

2 Investigation: Peel the pomelo 

a Collect a pomelo or grapefruit and a bucket of water.

b Put the pomelo in the bucket of water and observe what happens.

c Peel the pomelo, put it back in the water and observe what happens.

d Write up a scientific report to say what you did and what you found out.

e Explain the two different results in your conclusion.

3 Investigation: Diver in a bottle

a Collect a large plastic bottle with a screw cap, a plastic pen top, some plasticine (or other weight) and a paper clip or piece of wire.

(continued over)

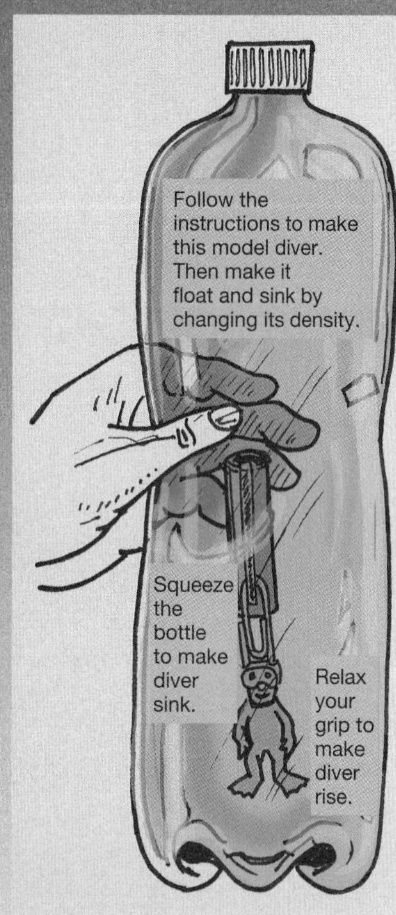

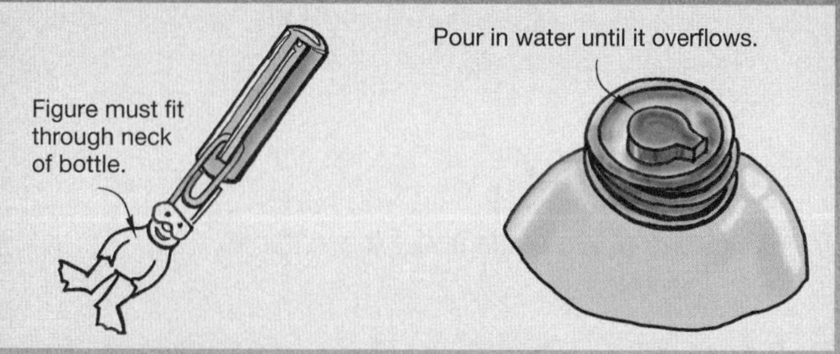

b Make a small plasticine model of a diver about 3.5 cm long. Fix the diver to the pen lid with the paper clip or piece of wire.

c Fill the bottle with water and put the lid and diver in the bottle.

d The pen top should float with its top just above the surface of the water. Make the model bigger or smaller if needed so that it will just float.

e Pour in more water so it is overflowing and screw on the top.

f Squeeze the bottle to make the diver sink and relax your grip to make the diver rise.

4 Investigation: Paper boats SR

a Collect several sheets of paper and some marbles or small stones.

b Lay a piece of paper in front of you. Fold it in half vertically to make a rectangle.

c With the creased edge at the top, fold down the top two outer corners of the rectangle, until the points meet. The paper should be in the shape of a triangle, with a slim rectangular base.

d Fold one side of the bottom edge up along the base of the triangle shape. Turn the boat over. Repeat on the other side.

e Then pull the sides gently apart at the bottom

f There are many different styles of boats. Different students can try making different styles.

g Try putting some marbles or small stones in each boat. Which boat can carry the most before it sinks?

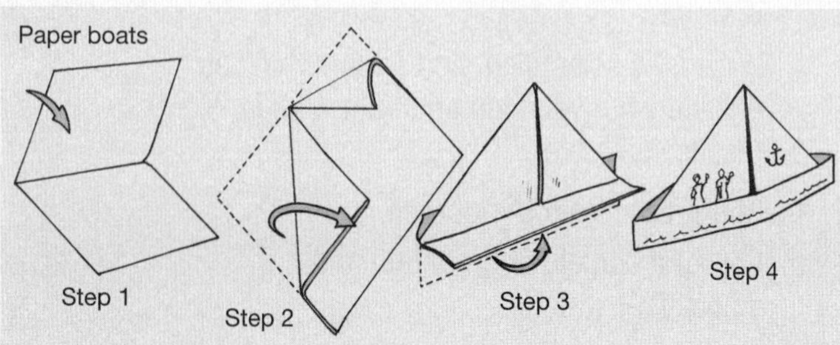

Investigating acids and bases

We have seen that materials or substances have physical properties such as strength, hardness and density. Substances also have **chemical properties** which can be useful but may be harmful. There are three important families of chemical compounds which are called **acids**, **bases** and salts.

Acids

Acids may be solids, liquid or gases although we usually use them when they are dissolved in water as solutions. All acids contain the element hydrogen and they all have a sour taste. The sour taste of unripe fruit is the typical taste of a dilute acid. Many foods contain weak acids which are harmless, for example:

- citrus fruits like oranges, lemons, pomelos and limes contain citric acid
- apples contain malic acid
- grapes contain tartaric acid
- tea contains tannic acid
- sour milk contains lactic acid produced by bacteria
- vinegar that is used to preserve food and add flavour contains acetic acid.

The air pollution from factories, power stations, trucks and cars can contain acids that are released as gases into the atmosphere. The gases can dissolve in rain drops, forming acid rain. Acid rain can hurt people's eyes, damage buildings, and kill plants and fish.

There are three other acids that are strong and must be used with great care: hydrochloric acid, sulphuric acid and nitric acid.

Strong acids are very **corrosive** and can cause serious burns to skin and make holes in clothes. If you get any strong acid on your skin or clothes you must wash it away immediately with plenty of water. The uses of common acids are shown in the table below:

Citrus fruits contain citric acid

Name of acid	Uses
Hydrochloric acid	Used to clean metals before soldering or plating with other metals. Dilute hydrochloric acid in the stomach helps to break down food and also kills germs.
Nitric acid	Used to makes explosives and dyes.
Sulphuric acid	Used in car batteries and to make the fertiliser called superphosphate.

All acids contain hydrogen which may be given off as a gas when they react with some metals. For example, acids will react with zinc, magnesium, iron and aluminium.

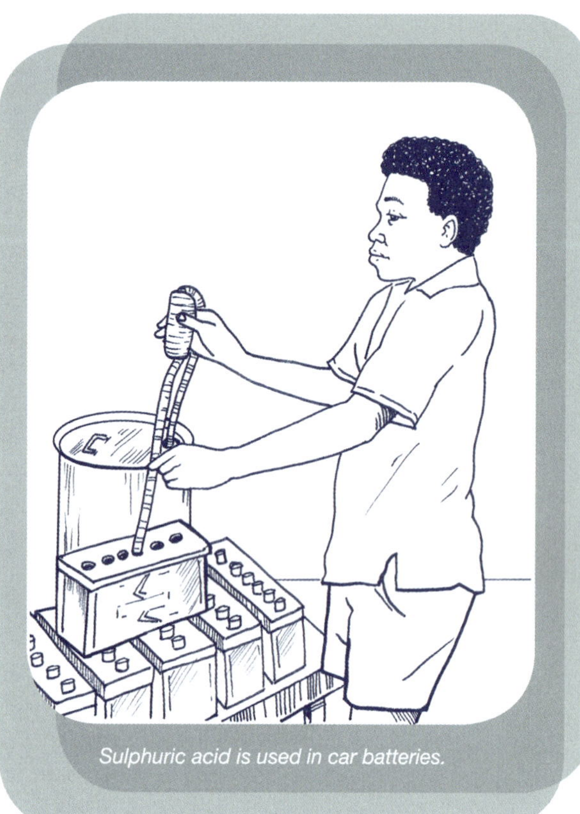

Sulphuric acid is used in car batteries.

Bases

Bases are chemical compounds that can be described as the opposites of acids. There are three common bases which have a number of uses: sodium hydroxide, ammonium hydroxide and calcium hydroxide. Bases that are soluble in water are known as **alkalis**. For example, sodium hydroxide is also known as caustic soda and is a strong base.

Strong bases are very corrosive and can cause serious burns to skin and make holes in clothes. If you get any strong base on your skin or clothes wash it away immediately with plenty of water. The uses of common bases are shown in the table below:

Name of base	Uses
Ammonium hydroxide	Used as cleaning agents that dissolve grease and dirt.
Calcium hydroxide	Used to remove hairs from animal skins when making leather, make paper from wood chips and in making cement and plaster.
Sodium hydroxide	Used to make soap, and to clean stoves and drains blocked with fat and grease.

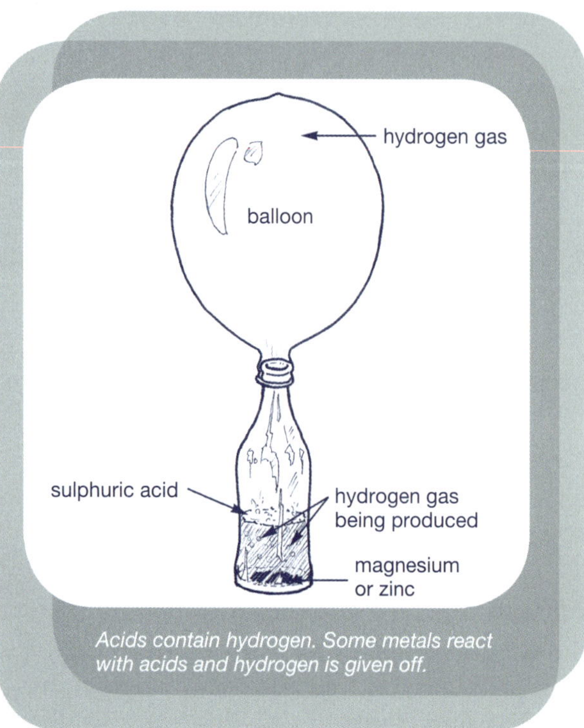

Acids contain hydrogen. Some metals react with acids and hydrogen is given off.

For you to try

1 Make a table of common substances that you use or find around your school or home and say whether each is an acid or a base. Compare your list of substances with other students and discuss any differences.

Some cleaning products that contain bases

Ammonia is the cleaning agent that dissolves grease.

Sodium hydroxide or caustic soda is used to clean drains.

Sodium hydroxide is used to make soap.

Indicators 

We use dyes to change the colour of grass skirts, pandanus mats and baskets, wool and other material. Some of these dyes will change colour when they are mixed with an acid or a base. Dyes like this are called **indicators**. We can test substances to find out if they contain an acid or a base by using an indicator. The most common indicator is called **litmus**. Litmus can be in the form of a solution or strip of paper that we dip into the solution to be tested. The way that litmus changes with an acid and a base is shown below:

With an acid	With a base
litmus indicator is red	litmus indicator is blue

When betel nut is chewed with lime the nut turns red. Betel nut contains a substance that changes colour in a similar way to litmus. This substance is colourless with acids, but with bases it turns red.

Many of the colours in flowers and vegetables are also good indicators. For example, we can remove the colour from red hibiscus flowers by boiling them for a few minutes. The solution can be used as an indicator, which is red with acids and green with bases.

For you to try

1 **Investigation: Making and using an indicator** SR A

a Collect a few red hibiscus flowers, a tin can, plastic lid, different substances such as juice from a lemon (or other citrus fruits), some rain water, some vinegar, soap, toothpaste, shampoo, different cleaning materials such as Ajax, Bon Ami, some ash from a fire (mixed with water) etc.

b Boil the red hibiscus flowers in a small amount of water in the tin can. Throw away the flowers but keep the red liquid—this is your indicator.

c Place a small amount of the substance to be tested on a plastic lid and add a few drops of the indicator and mix the two together. Observe what happens.

d Record your results on a table.

e Write up a scientific report to say what you did and what you found out.

f What conclusions can you make about the substances that you tested? Which were acids and which were bases?

2 List four properties of acids.

3 List four properties of bases.

The pH scale 

Universal indicator is a mixture of many indicators that shows if an acid or base is weak or strong. Scientists also use a scale of numbers called the **pH scale** and the numbers range from 0 to 14. The colours of universal indicator and the pH scale are shown in the table opposite.

The pH scale is used in everyday life. For example, gardeners and farmers are interested in the pH of soils because different crops grow best in different soils. For example:

- bananas, sugar and coconuts grow best in a slightly basic soil of pH 7 to 8
- peanuts, sweet potato and coffee prefer a fairly acidic soil of pH 5 to 6. After a bushfire the ashes will make the soil slightly basic and so the first plants to appear in the bare ground are

usually those that can grow in soil with a pH greater than 7. Later on the soil will become less basic and other plants will start to grow.

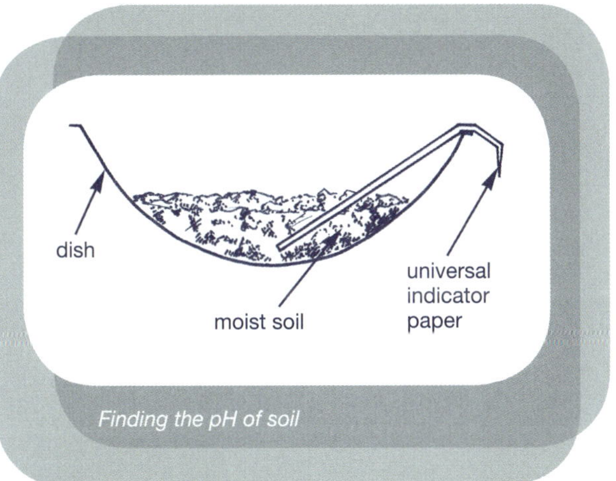

Finding the pH of soil

pH range	Colour of universal indicator	Strong/weak/acid/ base
0	red	
1	red	strong acids
2	red	e.g. hydrochloric acid
3	pink	
4	pink	weak acids
5	orange	e.g. citric acid
6	yellow	
7	green	neutral e.g. pure water, salt solution
8	blue-green	
9	blue	weak bases
10	blue	e.g. lime water
11	purple	
12	purple	strong bases
13	purple	e.g. sodium hydroxide
14	purple	

Lime being spread on soil that is too acidic

We can test the pH of the soil using universal indicator. If soil is too acidic farmers and gardeners can add lime to neutralise the acid and raise the pH of the soil.

Neutralisation

If an acid reacts with a base it can form a neutral solution containing a salt and water. Because they are chemical opposites the base will cancel out the effect of the acid. This is known as **neutralisation** and heat energy is also produced. The neutralisation reaction can be shown by the general equation, for example:

acid + base → salt + water + heat

hydrochloric acid + sodium hydroxide → sodium chloride + water + heat

Neutralising acids is important in everyday life. For example, sugar in your mouth produces acids that rot your teeth. Toothpaste is basic and neutralises these acids so it helps to clean your teeth soon after eating. Indigestion is often caused by having too much acid in the stomach.

Medicine containing the weak base magnesium hydroxide can neutralise the acid and take away the pain.

Too much stomach acid causes indigestion but this can be neutralised with medicine containing magnesium hydroxide.

For you to try

1. **Give one example of**
   - **a** a strong acid
   - **b** a weak acid
   - **c** a neutral solution
   - **d** a weak base
   - **e** a strong base
2. Why do gardeners and farmers sometimes add lime to soil?
3. Explain what is meant by neutralisation and give some examples from everyday life.

Science in the village: Making lime (PD) (SS)

Lime is often used when chewing betel nut in Papua New Guinea and is traditionally made by heating coral or shells over a fire for a long time. Coral and shells are made of calcium carbonate and as they are heated they change into lime. The chemical name of lime is calcium oxide. We can show this reaction with the following word equation:

coral or shells → lime + carbon dioxide

How much lime do we get? People who make lime in the village know that the mass of the lime produced is much less than the coral or shells that they started with. For example, if we start with one kilogram of coral or shells we would get just over half a kilogram of lime:

calcium carbonate	→	calcium oxide	+	carbon dioxide
1000g	→	560g	+	440g

The missing matter is the gas carbon dioxide. If we collected and weighed the gas we would find that it had a mass of 440g in this example.

Sharing betel nut is important among friends and during ceremonies and celebrations. However, regular chewing of betel nut with lime is a health risk and can lead to gum disease and mouth cancer.

Betel nut can be dangerous if consumed in large amounts.

Lime is made by heating coral or shells over a fire.

For you to try

Project: Make your own yogurt ML

a Collect a pot or saucepan, a vacuum flask, two bowls of different sizes, 250mL of long life or UHT milk and a small amount of plain yogurt. You will also need to be able to use a refrigerator.

b Heat the long life milk in a pot until it starts to boil, them remove from the heat.

c Warm the vacuum flask by adding a little boiling water and emptying it again.

d Stir two teaspoons of plain yogurt into the milk, pour the mixture in the flask and screw on the lid.

e Leave the mixture for seven hours so the bacteria can make lactic acid.

f Pour the yogurt into a small bowl. Stand the bowl in a basin of cold water and keep stirring the yogurt so that is cools quickly. This will stop the bacteria from making more lactic acid.

g Cover the bowl and put in a refrigerator. Leave it to get thick for about four hours.

h Yogurt is good to eat on its own or with fresh fruit.

Using energy in the home

Heat

Anything that can make something else move has energy. Energy is needed to make things work. We know that work is being done when things move. Heat is a form of energy that can do work. When we light a fire the heat energy that is produced makes the smoke and ashes move up into the sky. Many machines like trucks, boats, aeroplanes and rockets burn fuel in their engines to make heat energy. The heat energy is used by the engine to make the machine move.

Heat and temperature 

Heat is a form of energy that can make things hot. When heat energy is given to an object it usually gets hotter. If heat is removed from an object it usually gets cooler.

Temperature is a measure of how hot something is. Hotter objects have a higher temperature than cooler objects. When heat energy is given to an object, its temperature usually rises, but it falls when heat is removed.

Temperature is measured on a **thermometer**. The units of temperature are **degrees Celsius**. For example, normal body temperature is 37 degrees Celsius which is written as 37°C. The temperature of the air is measured and given as part of the weather report because it is something that affects us every day. Some examples of the air temperature in Papua New Guinea are shown in the table below:

Air temperature	What does it feel like?
0°C	Very cold night in the highlands—frost may occur.
10°C	Cold morning in the highlands.
25°C	Cool morning on the coast. Warm day in the highlands.
30°C	Hot day on the coast—quite common.
35°C	Very hot day on the coast—unusual.

Uses of heat

Apart from the use of heat in engines to make machines move, there are many other uses of heat. In all of these examples the heat must be controlled to make sure that the right amount of heat is used:

- cooking using fires and stoves
- drying things—like copra and coffee beans
- keeping people warm—sitting by the fire
- cleaning things—hot water is used for washing
- sterilising—a lot of heat kills germs
- melting and **casting**—liquid metal can be poured into **moulds** to get different shapes

- moulding and bending—heating metals and some plastics can make them softer so that they can be moulded and shaped
- treating injuries—heat treatment helps injured muscles
- producing light—the wire **filament** in a light globe gets so hot that it glows and gives off light
- making different substances—for example, heat is used to make glass.

Problems of heat

Although heat can be very useful, it can also have disadvantages. For example:

- Too much heat can make people feel uncomfortable and may make them sick or die.
- Heat can cause serious burns to the skin and muscles.
- Too much heat can start a fire which can do a lot of damage.
- Heat caused by friction is a problem in many machines. Lubricants like oil and grease can help to reduce the friction and moving air or moving water is often used to cool the machine.
- Some industries pump a lot of heated water into rivers and the sea that affects the plants and animals living in the water.

The movement of heat

Heat moves from places where the temperature is higher to places where the temperature is lower. Heat can move in three ways—by **conduction**, **convection** and **radiation**. The way in which heat moves depends on whether the material is a solid, liquid or gas.

Conduction

- Conduction is the main way in which heat moves through a solid. The heat gradually spreads out through the solid from the hot part to the cold part. Cold things feel cold because the heat is conducted away from your hand.
- Substances that conduct heat well are called good conductors. Most metals are good conductors and copper is one of the best, which is why copper pipes are used in **solar** hot water systems.
- Substances that do not conduct heat well are called poor conductors or insulators. Examples of good insulators are thick woollen clothes and blankets, plastics, cork, wood and glass.
- Plastics that do not melt are used as the handles of saucepans, kettles, electric frypans and clothes irons. These appliances can be picked up without heat being conducted through the handle to the hand.
- A plastic called polystyrene is used as an insulator in the walls of eskies and refrigerators to keep food and drinks cold.

An esky being used to keep ice-blocks cold

Conduction and the particle theory

Scientists explain the conduction of heat using the particle theory. Everything is made of particles that are constantly moving. When something is heated the particles vibrate faster. The vibrating particles are not able to move through a solid but the movement is passed on to the next particles. In this way the heat vibrations of the particles are spread throughout the solid.

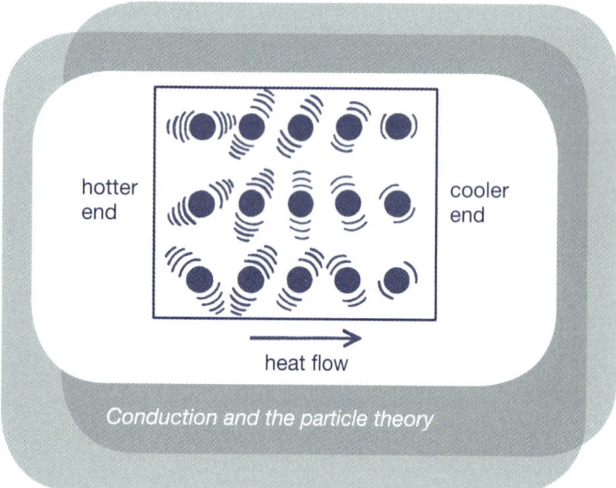

Conduction and the particle theory

How convection currents help to keep a coastal village house cool

Convection

- When air is heated it expands and floats upwards. Cooler air moves in to take its place. This movement of heat is called convection.
- Land and sea breezes are examples of convection currents in the atmosphere. Big convection currents in the Earth's atmosphere help to make winds.
- Convection also occurs in chimneys and house ventilation.
- Convection currents also occur in liquids. The hotter part of a liquid expands, becomes lighter and is pushed to the top. The colder, heavier liquid then moves in to take its place.
- Convection currents can be seen in hot drinks and also occur in the sea.
- Convection is used in solar hotwater systems and some engine cooling systems.

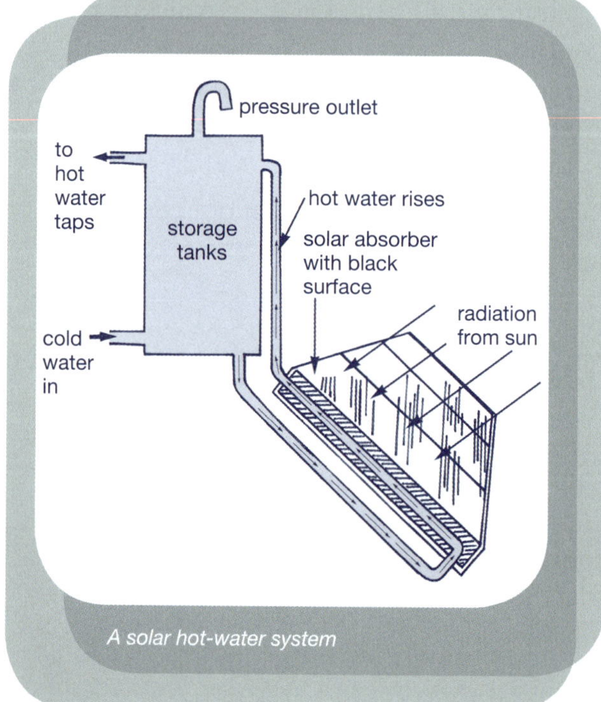

A solar hot-water system

Radiation

There is no air in the space between the Earth and the Sun and so the heat from the Sun cannot reach the Earth by conduction or convection. Both of these methods of heat movement need particles that can vibrate to carry the heat. We can still get heat from the Sun because it gives out waves of **electromagnetic radiation** such as light and infra-red radiation. An object gets hot when it absorbs radiation but the radiation itself is not hot. Everything gives off some radiation. The hotter it is, the more it radiates. Dark coloured objects and rough surfaces radiate more than light coloured and smooth surfaces.

Motorcycles and lawn mowers have cooling fins on the outside of the engine. These fins increase the surface area which radiates heat and so help cool the engine.

Dark coloured surfaces also absorb more radiation than light-coloured surfaces. For example:

- you feel hotter wearing dark clothes in sunlight than light coloured clothes
- black cars get hotter than white cars
- the copper pipes of solar hot water systems are painted black so they can absorb as much of the Sun's radiation as possible.

Light coloured and shiny surfaces reflect most of the radiation that falls on them. This is why sheets of shiny material made from aluminium are put under the roofs of schools and other buildings.

The vacuum flask

The vacuum or thermos flask is a good example of the way we use our knowledge of conduction, convection and radiation.

- The **vacuum flask** is a bottle with a double wall usually made of glass which is a poor conductor. The outside case is made of metal or plastic for protection.
- The inner surfaces are shiny to prevent radiation.
- Air is removed from the space between the walls to make a vacuum which prevents conduction and convection.
- The stopper is made from a poor conductor such as cork, plastic or rubber.
- By reducing conduction, convection and radiation in this way, liquids can be kept hot or cold over a long period of time.

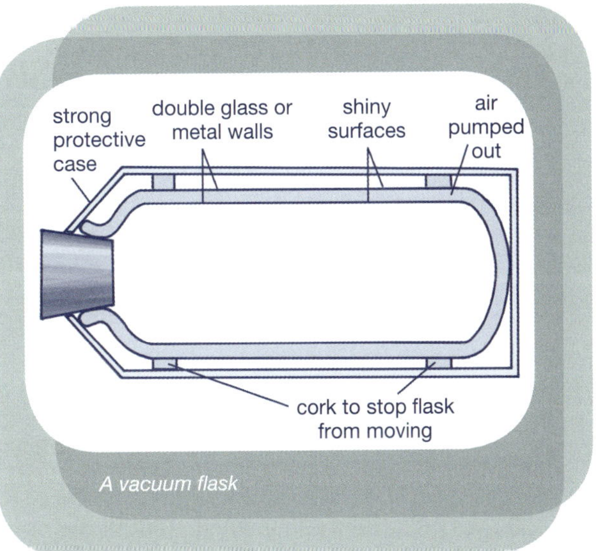

A vacuum flask

For you to try

1 Investigation: Conduction in solids

a Collect a small cooking pot and some different solids such as a 10 cm nail, a piece of fencing wire, a metal spoon, a plastic spoon, a wooden stick, an old biro case, the strap from an old rubber thong and a glass bottle.

b Half fill the pot with water and heat the water.

(continued over)

The temperature of the water should be about the same as hot tea, but not boiling.

c Put each of the different solids in the hot water and touch the end to find out if it gets hot.

d Write up a scientific report of your investigation to say what you did and what you found out. Include a table of results and decide which solids are good conductors and which are poor conductors.

2 **Investigation: Convection in liquids** SR

a Collect a clear glass jar, hot water and something to colour the water-like ink, dye, a little instant coffee powder, or the red colour from betel nut that has been mixed with lime.

b Almost fill the jar with hot water. The temperature of the water should be about the same as hot tea, but not boiling.

c The jar needs to be in a place where you can see it easily but where it will not be disturbed or get knocked over.

d Very carefully add a small amount of the colour to the liquid at the side of the container.

e Observe what happens in the water.

f Write up a scientific report of your investigation to say what you did and what you found out.

3 Describe some examples where insulation is used in a house in order to keep it cool. L

4 With a diagram, describe how a chimney on a stove uses convection currents. L

5 Explain why people making a mumu need to use insulators. What insulators do they use? L

6 What colour would you choose for a car in a place where the days are hot and sunny? Give reasons for your choice.

7 Which would melt more quickly on a sunny day—a plain white ice-cream or a chocolate flavoured ice-cream? Explain your answer.

8 Explain why a vacuum flask can be used to keep hot liquids hot and cold liquids cold. L

9 Use the particle theory to explain why conduction of heat can occur in solids but is not really possible in gases.

Expansion

When most substances are heated they become slightly larger. This increase is called **expansion**. When the same substances are cooled they get smaller. This decrease in size is called **contraction**. The expansion of solids is usually so small that it is difficult to see, although some solids expand more than others. Liquids usually expand faster than solids and gases expand faster than liquids. The following are examples of expansion:

- A corrugated iron roof will make creaking noises as it expands in the hot sun and again when it cools down and contracts.
- Bridges must have small gaps to allow them to expand on a hot day without cracking.
- Overhead power lines increase in length and hang lower in hot weather because they expand.
- Electric irons, electric frypans and electric stoves are controlled by a special switch called a **thermostat**. The thermostat turns on and off because of the expansion and contraction of a metal strip inside the switch.

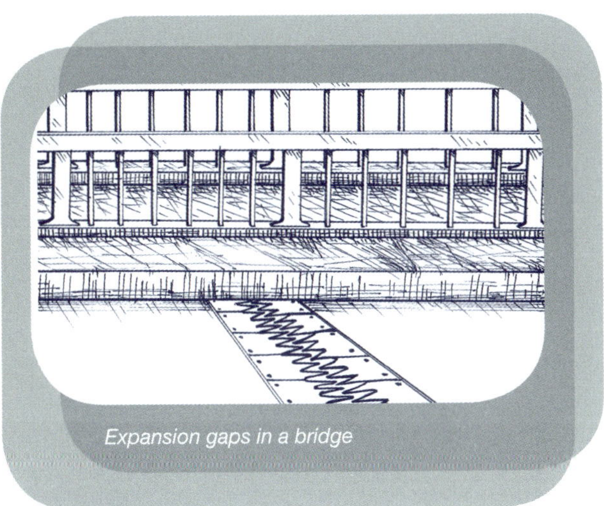
Expansion gaps in a bridge

Electricity

Electricity is a form of energy that is used every day in many things such as torches, radio cassette players, fans, refrigerators and stoves. Some of these appliances use batteries while others plug into the mains supply in houses, schools, offices and factories. Electricity that is used in buildings must come from another source of energy, such as a diesel powered generator, **hydro-electricity** or solar panels that use energy from the Sun.

Batteries or dry cells like those used in torches are a safe and convenient way to store electricity. These dry cells have two places where the wires of a piece of electrical equipment can be connected. They are called the **terminals** of the cell and are shown in the diagram below:

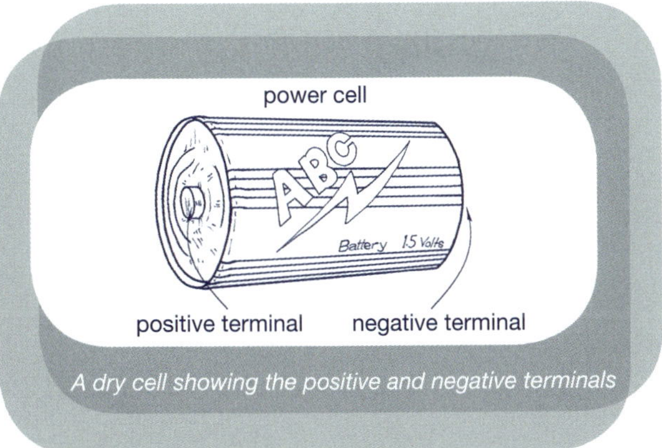

A dry cell showing the positive and negative terminals

A **circuit** is a path around which electricity can flow and is usually made of wires. When we turn on a torch the electricity or electric current flows through the cells, round the circuit through the bulb and back to the cell again. Wire and other materials that allow electricity to flow in this way are called conductors. Materials that will not allow electricity to pass through them are called insulators. Examples of conductors and insulators are shown in the table below:

Conductors	Insulators
silver	glass
copper	plastic
aluminium	air
iron	brick
graphite—a form of carbon	dry wood
salt water	pure water

Scientists and electricians use electrical symbols to draw circuit diagrams. The same symbols are used all over the world which makes it easier for everyone to understand a circuit. The diagram below shows a picture of a circuit with two dry cells, a switch and a bulb, and a circuit diagram using electrical symbols.

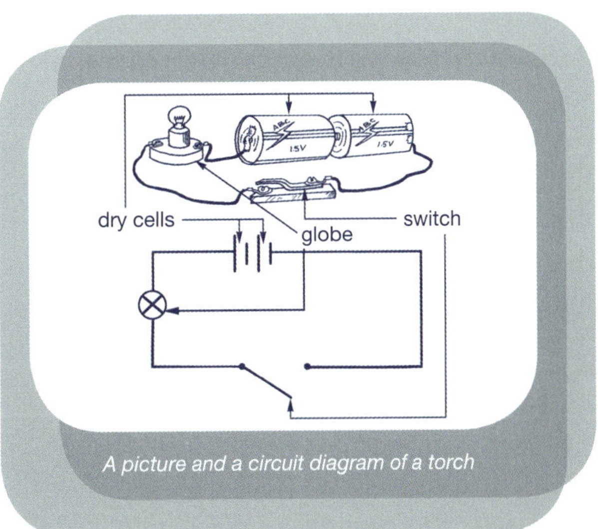

A picture and a circuit diagram of a torch

Series and parallel

There are two different ways of connecting bulbs or other parts into an electrical circuit. They can be connected in series or parallel.

Series circuits are the simplest circuits and everything is connected in a line or single loop. When we put the dry cells in a torch or radio they are connected in series. When we connect two or more bulbs in series they become dimmer because there is less electric current in the wires than before. If one bulb is not working or if a wire is not connected properly, the whole circuit stops working.

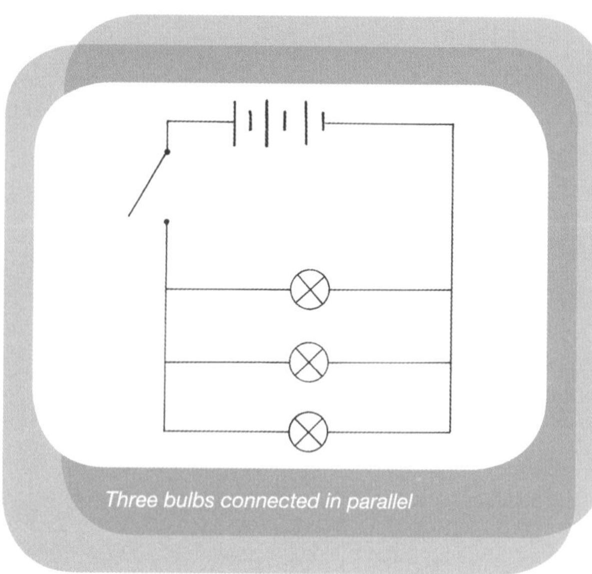

Three bulbs connected in parallel

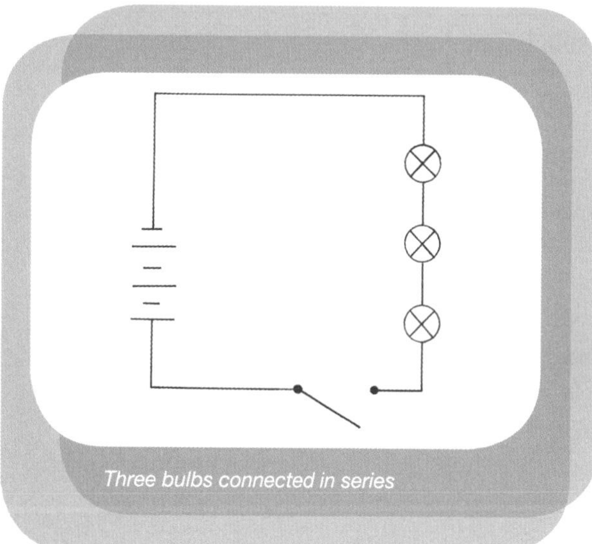

Three bulbs connected in series

Parallel circuits are made from several loops and the current divides and flows to each part separately as shown in the diagram at the top of the page.

When we connect two or more bulbs in parallel they do not become dimmer because each bulb receives the same electric current. However, two bulbs will take more current than one bulb, so the battery will not last as long. If one bulb is not working or is turned off there is still a complete circuit through the other bulbs which will still work. Parallel circuits are more useful because each part can be controlled separately.

Parallel circuits are used in house wiring. The main switch at the meter box turns the power on. When the main switch is off no electricity can flow to any part of the house. The switches inside a house are used to control the lights and power points independently. A simplified part of the wiring inside a house is shown in the diagram below.

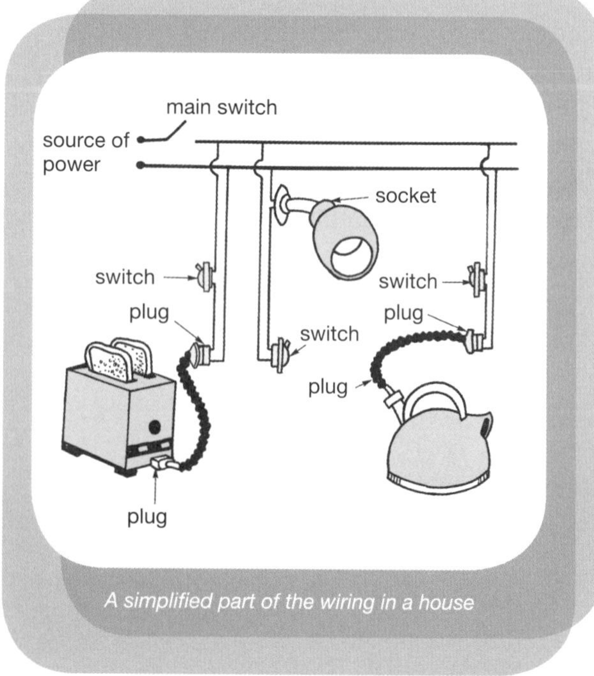

A simplified part of the wiring in a house

For you to try

1 **Investigation: Conductors and insulators** 

a Collect two dry cells, a bulb and holder, several pieces of wire and some materials to test: metal objects such as a nail, an aluminium can and a piece of copper pipe, a glass marble, plastic, wood, stone, brick, paper etc.

b Make a simple circuit as shown in the diagram on page 83 for a torch.

c Use the circuit and test each material to see if it will conduct electricity.

d Write up a scientific report to say what you did and what you found out. Include a circuit diagram in your method and a table of results. What conclusions can you make about the materials that are conductors?

2 **Investigation: Series and parallel**  

a Collect two dry cells, three bulbs and holders, and eight pieces of wire.

b Use the circuit diagrams on page 84.

c Set up a simple circuit with one bulb. Observe what happens to the brightness when you have two bulbs in series and then three bulbs in series.

d Set up a simple circuit with one bulb. Observe what happens to the brightness when you have two bulbs in parallel and then three bulbs in parallel.

e Write up a scientific report to say what you did and what you found out. Include circuit diagrams in your method and a table of results. What conclusions can you make about the brightness of two or more bulbs when they are connected in series and in parallel?

3 Explain why screwdrivers and pliers used by electricians have plastic handles and plastic covering some of the metal parts. 

4 a Explain why electrical appliances that we use in the home usually have at least two wires to make them work.

b Many cars and trucks have a chassis or frame that is made of steel. The negative terminal of the battery is also connected to the chassis. The horn on a car or truck can be controlled by having only one wire from the battery and then a short wire from the horn to the chassis or frame. Draw a simple circuit diagram and explain how a complete circuit can be made so that the horn works.

Uses of electricity

The different ways that we use electricity can be divided into three main groups or effects. The electrical appliances that people use depend on the fact that electricity can produce heat, light or a magnetic effect.

The mining industry also uses large quantities of electricity to obtain copper and aluminium from metal ores that come from rocks in the ground. Two metal plates called **electrodes** are dipped into a liquid that contains the crushed ore and connected to an electric current. Chemical reactions can be seen at the two plates and pure metal builds up on one of the plates. This reaction is known as the chemical effect of an electric current.

The heating effect

Most metals are good conductors but some mixtures of metals called **alloys** do not conduct electricity so easily. For example, copper wire is a much better conductor than nichrome wire which is a mixture of nickel and chromium. Because wires made from these alloys do not let electricity pass easily we say that the wire has more resistance to electricity. The thickness and length of the wire also affects the resistance, as shown in the diagram below:

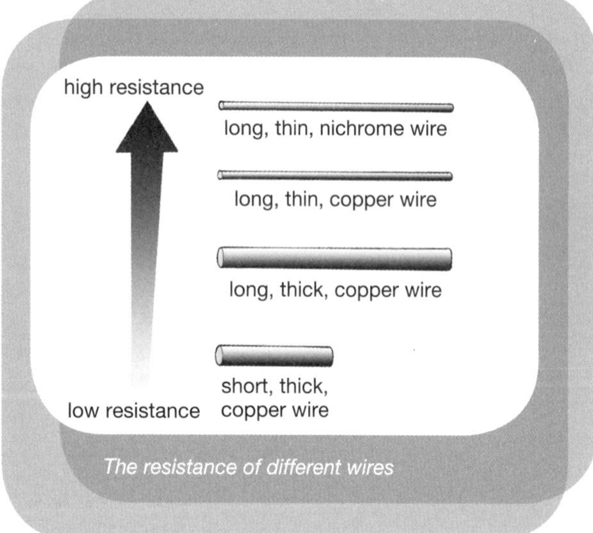

The resistance of different wires

Whenever electricity flows through a resistance wire, heat is given off because energy has to be used to overcome the resistance. This idea is used in the heating elements of appliances such as electric jugs, electric irons, frying pans and stove elements. The resistance wire is usually wound in a coil called an element.

Electrical appliances that get hot must be used carefully so as not to burn out the element and start a fire. For example:

- Electric jugs and urns must have water completely covering the element before they are switched on.
- An electric shower heater must have water flowing over the element before the power is switched on.
- An electric iron must be stood on its heel when it is not being used.

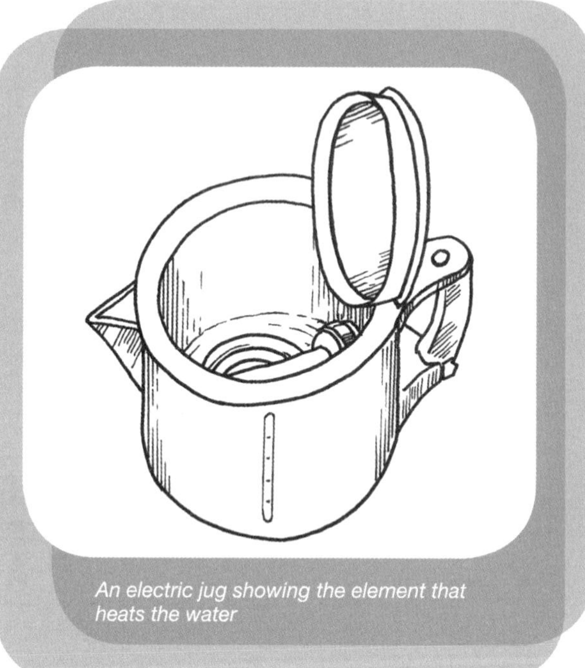
An electric jug showing the element that heats the water

When using a shower heater, the water must be flowing before the power is turned on.

The lighting effect

Electrical energy can be converted into light energy in two main ways:

Light bulbs

- Light bulbs contain a coil of very thin, high resistance wire called a filament.
- The filament is made of a metal called tungsten that has a melting point of 3500°C. When electricity passes through the filament it becomes very hot and gives off light.
- Most bulbs are filled with the gases nitrogen and argon that help prevent the element from melting or burning.

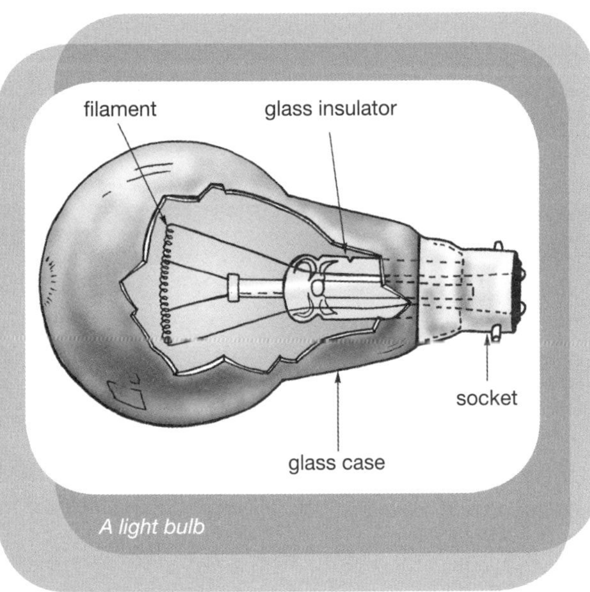

A light bulb

Fluorescent tubes

- Fluorescent tubes contain a special chemical powder coating on the inside of the glass.
- The electric current produces ultraviolet radiation which causes the chemical to produce light.
- Fluorescent tubes are about three times as efficient as light bulbs and they cast little shadow because they are long.

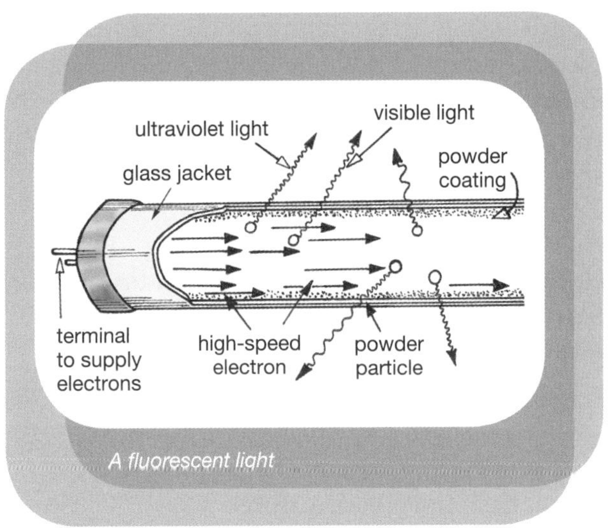

A fluorescent light

The magnetic effect

When an electric current passes along a wire there is a magnetic effect. If a battery is connected to a wire and a small compass is put near the wire the compass needle moves. This shows that a magnetic field is produced around the wire.

The magnetic effect can be increased by using a coil of wire instead of a straight piece of wire. If an iron bar is put inside the coil the effect is even greater. An iron bar inside a coil of wire is called an **electromagnet** and only works while the current is flowing.

Large electromagnets are used to lift iron and steel in steelworks and scrap metal yards. Electric bells, buzzers and car signal lights also use small electromagnets to make them work. The electromagnetic effect is also used in the record and playback heads of tape recorders, video cassette recorders, in loudspeakers and in television screens.

The electric motor

The most important use of the electromagnetic effect is in electric motors which combine electricity and magnetism to produce movement. Electric motors of different sizes are used in many different ways. Examples include radio cassette players, fans, washing machines, power tools and starter motors in cars and trucks.

An electric motor consists of a rotating electromagnet called the **armature** and two fixed magnetic poles, which may be permanent magnets or electromagnets. When a current is passed through the armature it becomes an electromagnet and interacts with the fixed magnetic poles. This causes the armature to rotate. Each time the armature rotates 180 degrees a device called a **commutator** reverses the direction of the current in the armature so that the armature continues to rotate in one direction.

The working of an electric motor can be summarised as follows:

Coil + Current + Magnet → Movement

The electric motor converts electrical energy to **kinetic energy** without producing pollution.

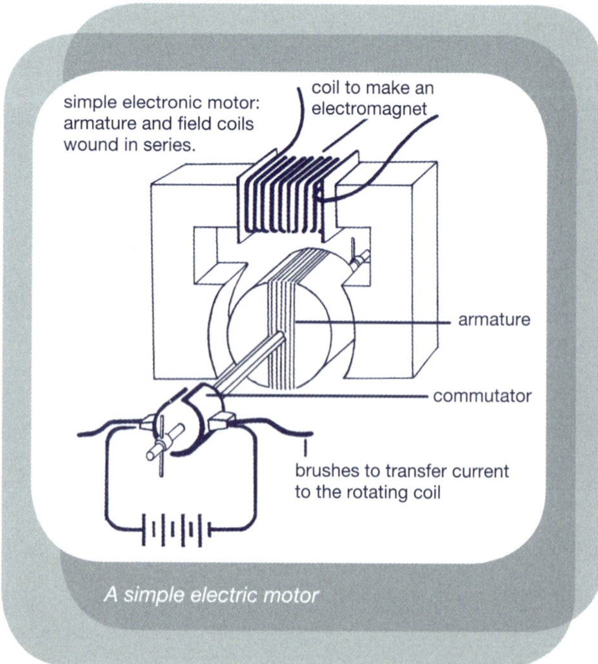

A simple electric motor

For you to try

1 Investigation: Make an electromagnet SR A

a Collect a large iron nail, a dry cell and about one metre of insulated wire and some paper clips.

b Strip about 2 cm of the plastic covering from both ends of the wire.

c Wind the insulated wire around the nail about 30 times.

d Hold one end of the wire on the positive terminal of the dry cell and the other end of the wire on the negative terminal.

e Hold one end of the nail near some paper clips and observe what happens.

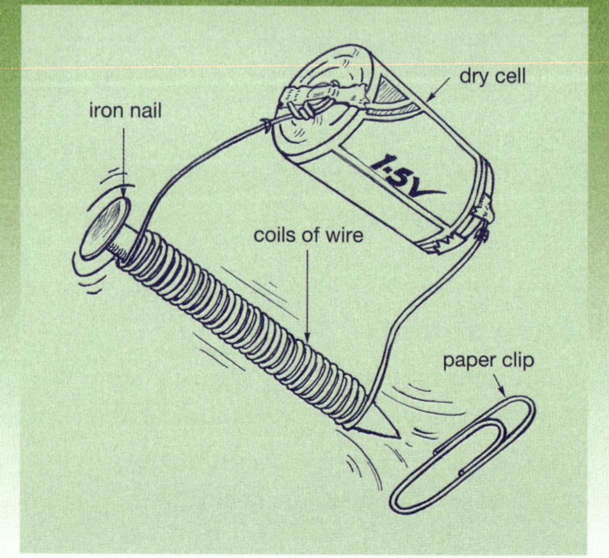

2 Investigation: Make a simple current meter

SR

a Collect a new dry cell, an old dry cell, about one metre of insulated wire, a matchbox tray, a small compass and some plasticine or sticky tape.

b Fix the small compass in the tray of a match box with plasticine or sticky tape so that it does not move.

c Strip about 2 cm of the plastic covering from both ends of the wire and wrap a few coils of wire around the box.

d Turn the match box tray so that the needle of the magnet points across the box in the same direction as the coils of wire, as shown in the diagram. Fix the match box tray in this position.

e Connect the wires to one of the dry cells and observe what happens. Do not leave the wires connected for too long because the battery will go flat very quickly.

f When the electricity is connected the compass needle is moved by the magnetic effect of the electric current. The amount that the compass needle moves tells you about the size of the current.

g Repeat the experiment with the other dry cell. What is the difference between the new dry cell and the old dry cell?

h Connect the wires to the opposite terminals of the battery, repeat the two experiments with both dry cells and observe what happens.

i Write up a scientific report of your experiment to say what you did and what you found out.

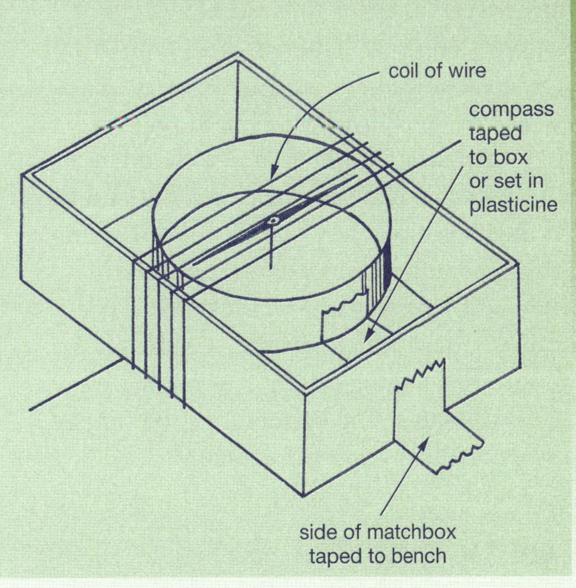

A simple electric current meter

3 Explain why thick copper wire is better than thin for use in connecting wires. Why would you not use nichrome wire for connecting wire?

L

Producing electricity

Just as electricity produces certain effects, so the effects of electricity can be reversed to produce electricity. For example, heat, light, magnetism and chemical reactions can all be used to make electricity.

Chemical methods

Cells use chemical reactions to make electricity. The most common type of cell is the dry cell which is used in torches and radios. In this type of cell the outside metal case is made of zinc and is known as the negative electrode. The zinc case is usually covered with the label.

The carbon rod in the middle of the cell is the positive electrode. There is a paste containing different chemicals between the two electrodes so that a chemical reaction takes place which produces electricity. One electrode is used up during the chemical reaction and then the battery is said to be 'flat' and is usually thrown away.

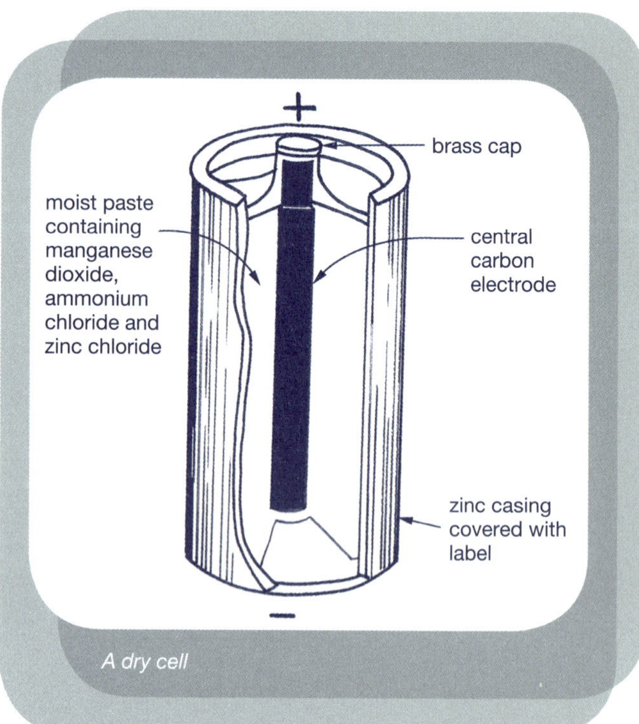

A dry cell

Another type of cell is the lead accumulator which is used in car batteries. This type of cell has a series of lead and lead-coated electrodes that are covered with dilute sulphuric acid. When the battery is being used a chemical reaction occurs between the electrodes and the acid that produces the electricity. An advantage of this type of battery is that it can be recharged by passing an electric current back into the battery which reverses the chemical reaction. Recharging produces some hydrogen and oxygen gas so the battery must be topped up with distilled water occasionally.

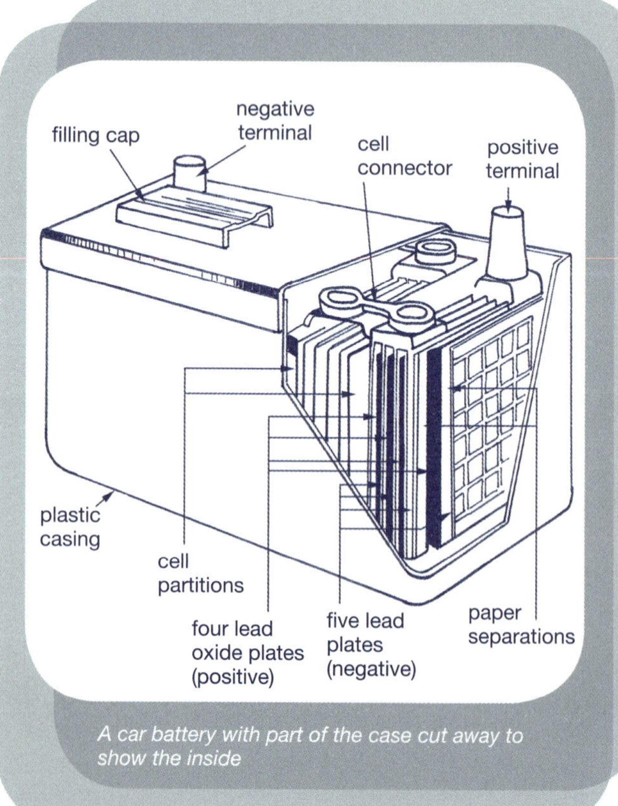

A car battery with part of the case cut away to show the inside

Magnetic method

When a **magnet** is moved inside a coil of wire, electricity will be produced in the coil. The kinetic or moving energy is changed to electrical energy. If there is no movement then the current will not flow. We can describe what happens simply in the following way:

Coil + Magnet + Movement → Electricity

This is the way that a generator or dynamo works. A magnet is made to spin inside a coil of wire. The spinning can happen in different ways.

- A bicycle dynamo is turned by one of the wheels which is turned by a person's leg muscles.
- A car generator is turned by the engine that burns petrol or diesel.

- Bigger generators like those in power stations are driven by diesel engines or turbines.
- A turbine is a special wheel that has blades that can be turned by a liquid or a gas. The turbine is then used to turn the generator. When water is used to turn the generator then hydro-electricity is produced.
- The bigger hydro-electric power stations are at Rouna, outside Port Moresby, at Yonki in the Eastern Highland and Warangoi in East New Britain, although there are also many smaller ones in other provinces.

A hydro-electric generator

Light method

Solar cells are special cells that convert sunlight directly into electrical energy. Solar cells are used to provide electricity for telecommunication repeater stations in the bush and for navigational aids for shipping.

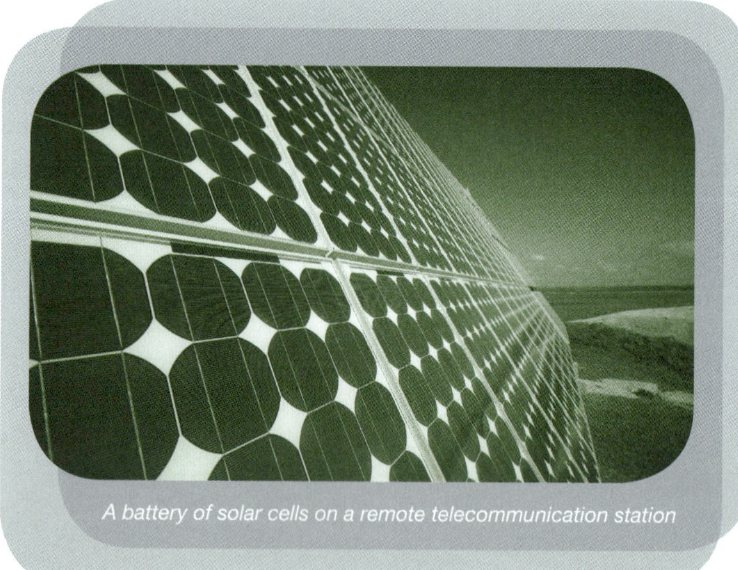

A battery of solar cells on a remote telecommunication station

Safety rules for using mains electricity

Mains electricity can be used in many ways that are helpful but it can also be very dangerous. It can cause fire and electric shock, so it is best to use it carefully. These safety rules should always be followed:

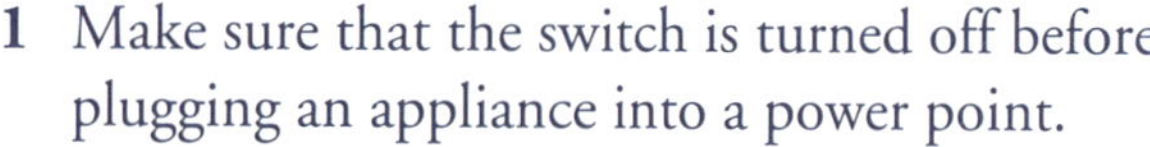

1 Make sure that the switch is turned off before plugging an appliance into a power point.
2 Switch off the power before you remove a plug from an appliance or power point.
3 Do not pull a plug out by the cord. Take hold of the plug itself.
4 Do not fill an electric jug or pour from it while the power is switched on.
5 Do not let wires or electrical appliances get wet. Do not touch switches or electrical appliances if you have wet hands or wet feet.
6 Do not have long extension leads in and around the house. Somebody could easily trip and knock the appliance over.
7 Do not use adaptors to run more than two appliances from a single power point.
8 Do not try to clean an appliance until you have switched it off and removed the plug.
9 Do not poke your fingers or push other objects into electrical appliance, power points or fuse boxes.
10 Do not use appliances with worn or damaged cords.
11 Keep well away from overhead wires when you are working or playing. Do not throw anything over the wires or fly kites near them.
12 Do not climb electricity poles or pylons.

Force, work and energy

Measuring force 

A force is a push or pull and is measured in units called newtons (N). The four main types of forces are:

- gravitational forces
- magnetic forces
- electrical forces
- elastic forces.

There is a small force between all things. For example, two people sitting next to each other are being pulled together by a very weak force. It is hard to understand this because the force is so small that we cannot feel it. However, when one of the objects is as large as the Earth, then the force is quite large. This force is called **gravity** and it has the following effects:

- holding things on the surface of the Earth
- making things fall to the ground when they are thrown in the air
- holding **satellites** and the Moon in orbit around the Earth.

The force of gravity pulling on an object is called weight. Weight can also be measured in newtons, like other forces. Forces can be measured with a force-meter, which is also called a spring balance. There is a spring inside the force-meter and the greater the force the more the spring stretches and the further the pointer moves along the scale. The Earth pulls a mass of 100 grams with a force that is very close to 1 newton (N). You can feel a force of 1 newton by holding a 100-gram packet of tea in your hand.

A one kilogram mass like a packet of sugar has a weight on the Earth of about ten newtons.

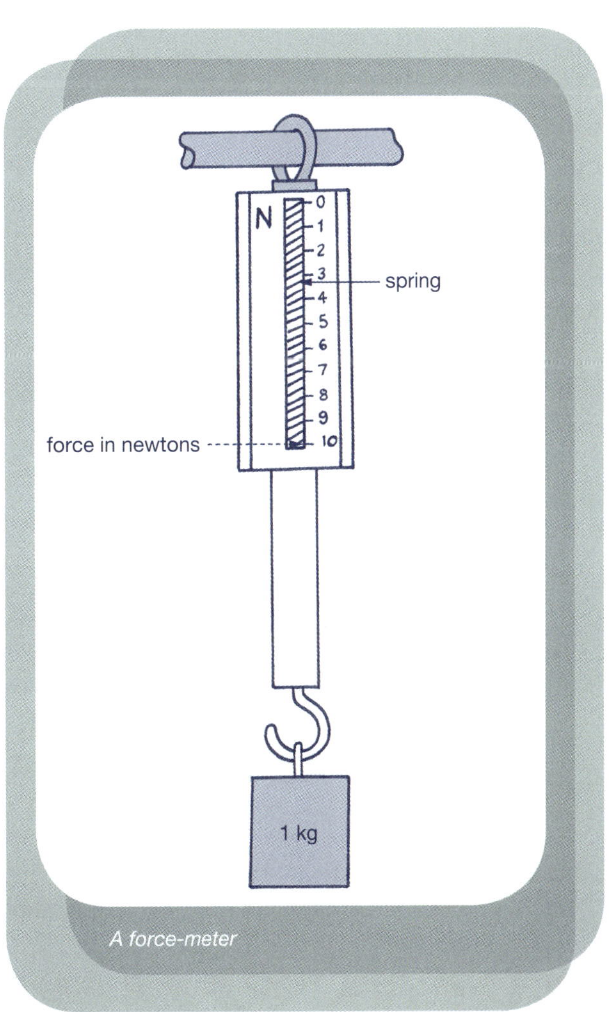

A force-meter

The table shows the weight force in newtons of some everyday masses.

Mass	Weight on the Earth
Packet of tea (100 grams)	1 newton
Packet of sugar (1 kilogram)	10 newtons
Bag of rice (5 kilograms)	50 newtons
Bag of cement (50 kilograms)	500 newtons

The force of gravity not only depends on the size of objects, it also depends on the distance an object is from the Earth. As we move further into **space** the force of gravity gets less.

If a 100-gram packet of tea was taken to the Moon it would not weigh 1 newton any more. It would weigh only about one sixth of a newton. It would still have a mass of 100 grams, but because the Moon is much smaller than the Earth the pull of gravity on the Moon is only one sixth of the gravity of the Earth.

Measuring work 

When energy is used work is done. Scientists use the word 'work' when a force moves an object a certain distance. If something is not moving then no work is being done. We can measure work in units called **joules (J)**. This unit is named after the British scientist James Joule who was born nearly 200 years ago. His father owned a brewery so he helped to make beer in the day and did his science experiments at home after work.

One joule is the amount of work done when a force of one newton moves an object a distance of one metre. If a force of two newtons moves an object a distance of two metres then four joules of work are done.

There is no instrument for measuring work, it can only be calculated. We use a force-meter to measure force in newtons, and a rule or tape to measure the distance in metres. We then multiply these two readings using the formula below and get an answer for work in joules:

work (J) = force (N) × distance (metres) (m)

Measuring energy

When work is done, energy is used. The unit in which energy is measured is also the joule. To do one joule of work you must have at least one joule of energy. However, because the joule is a very small unit it is sometimes more convenient to use bigger units like **kilojoules (kJ)** or **megajoules (mJ)**. One kilojoule is a thousand joules and one megajoule is a million joules.

For you to try

1 Investigation: Measuring forces SR A

a Collect a thin strip of timber about 30 cm long, five D cells, a plastic bag, a piece of wire and a rubber band.

b Make a small groove at the top of the strip of timber.

c Put the rubber band in the groove and put one D cell in the plastic bag and hang it on the rubber band.

d Observe how far the rubber band stretches when one D cell hangs from the rubber band. This shows the force (1 N) needed to pull an object weighing 100 g since a D cell weighs about 100 g. Mark this position with the number '1' on the timber. Remember 100 g = 1 N.

e Add another D cell to the rubber band, observe how far it stretches and again mark the position on the timber.

f Add more D cells and complete the scale up to 5 N. This is called **calibrating** the force-meter.

g Now use the force-meter to measure the size of the following forces:

- force needed to pull the door open
- force needed to drag an empty bilum or pencil case across the floor
- force needed to open a drawer
- force needed to open the cover of a book.

h Write up your experiment to say what you did and what you found out. Include a table of results.

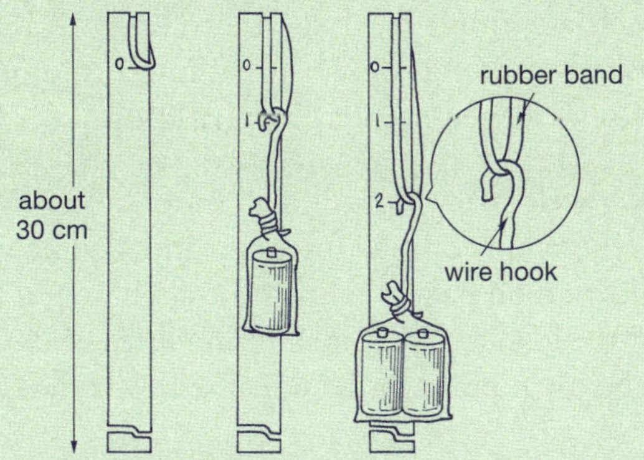

Calibrating the force-meter

2 Investigation: Measuring friction SR

a Collect a block of wood, a nail, a book, 30 marbles, some sand and the force-meter from the previous investigation.

b Place the block of wood on the floor and move it using the force-meter as shown in the diagram below. Measure the force needed to make it move.

c Place the block of wood on the marbles and again measure the force needed to make it move.

d Place the block of wood on the floor, put the book on the block of wood and again measure the force needed to make it move.

e Spread some sand on the floor, place the block of wood on the sand and measure the force needed to make it move.

f Write up your experiment to say what you did and what you found out. Include a table of results.

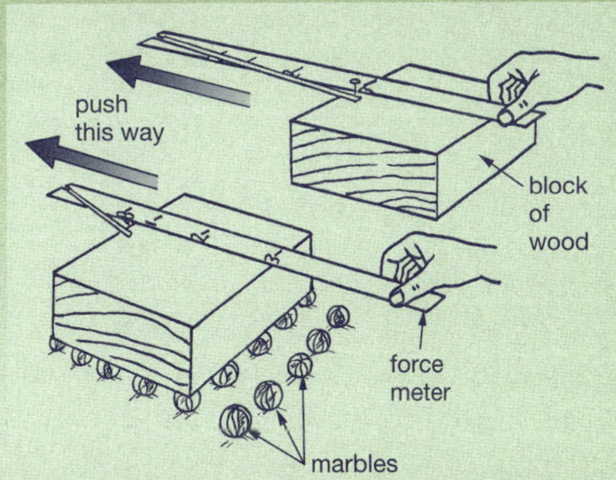

Measuring the friction of blocks of wood

3 Copy and complete the following table to show which of the actions involve a push force, pull force or no force. A

Action	Push, pull or no force
Opening a window	
Turning a screw with a screwdriver	

4 List the four main types of forces. Give an example of the use for each. 

5 At the surface of the Earth, what is the approximate weight in newtons of the following masses?

a 100 grams

b 600 grams

c 250 grams

d 1 kilogram

e 1.8 kilograms

6 What is the reading on each of the force-meters in the diagram below? 

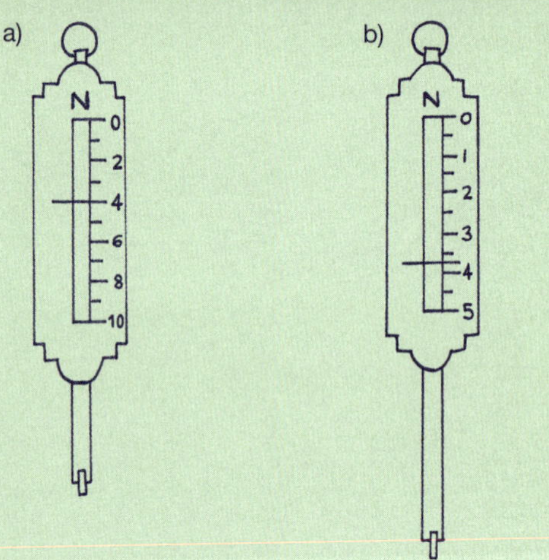

7 Write down four things that you have done today and describe how you used energy while you did these things.

8 Describe the relationship between work and energy.

9 How many joules of energy are used in each of the following cases?

a 100 grams is lifted up 1 metre

b 200 grams is lifted up 1 metre

c 200 grams is lifted up 5 metres

d 1 kilogram is lifted up 1 metre

Simple machines

A simple machine is something that helps to make work easier. For example, it is easier to move a bag of cement or a load of rocks if you put them in a wheelbarrow than if you try to carry them in your arms. We say that the wheelbarrow is a simple machine.

The main types of simple machines are:

- levers
- the wheel and axle
- inclined planes
- pulleys and gear wheels.

Levers

Cars and trucks have a gear lever in order to change gear. When we get a puncture in a car or a truck, we need a tyre lever to remove the tyre from the wheel rim. Other tools which work in the same way are a crowbar, a pinch bar, a spade, a wheelbarrow and a bottle opener. All of these tools are simple machines called levers. A lever is usually a strong, rigid bar that can turn or **pivot** around a fixed point called a **fulcrum**.

A wheelbarrow lets us lift a heavy load easily because only a small force is required. The diagram on page 97 shows the large weight force of the load and the small lifting force of the man. The pivot point in the wheelbarrow is the axle.

The easiest way to use a wheelbarrow is to put the load close to the pivot point and to have your lifting force, or effort, far from the pivot.

- The distance between the load and the pivot is called the load distance.
- The distance between the effort and the pivot is called the effort distance.
- When the effort distance is larger than the load distance then there is a force advantage.

The wheelbarrow is a simple lever in which there is a force advantage.

- This means a *small* force is used to move a *large* load. We can redraw the diagram of the wheelbarrow to show effort distance and load distance. To do this we leave out the person, the load of soil and the wheel and in their place we put arrows to represent the load and the effort, as shown in the diagram below.

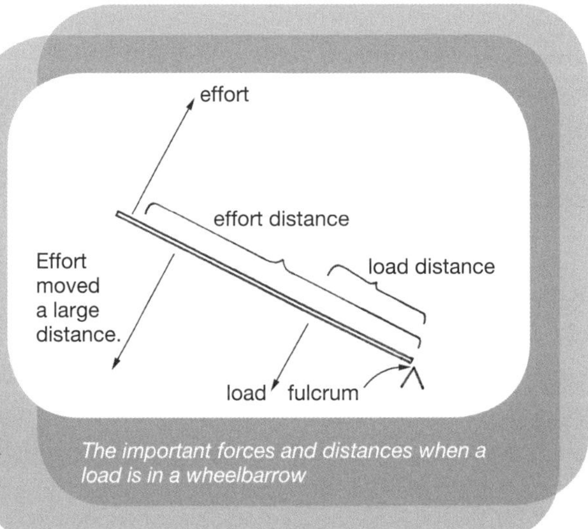

The important forces and distances when a load is in a wheelbarrow

The **mechanical advantage** of a simple machine compares the size of the effort or lifting force with the size of the load or weight force. In other words, the mechanical advantage is a force ratio which we can calculate by dividing the load by the effort as follows:

$$\text{mechanical advantage} = \frac{\text{load}}{\text{effort}}$$

Some simple machines are made of double levers. The bamboo tongs that are used for lifting hot mumu stones or turning the food cooking in a fire are an example of a double lever. Other examples are scissors, pliers, tin snips and forceps or tweezers.

The wheel and axle

A simple machine that uses the principle of the lever is the wheel and axle, which is made of two wheels that are joined together. The smaller wheel is called the axle. The screwdriver and the steering wheel of a car or truck are both examples of the wheel and axle.

To gain a force advantage, a small force is used to turn the larger wheel. This causes a large force

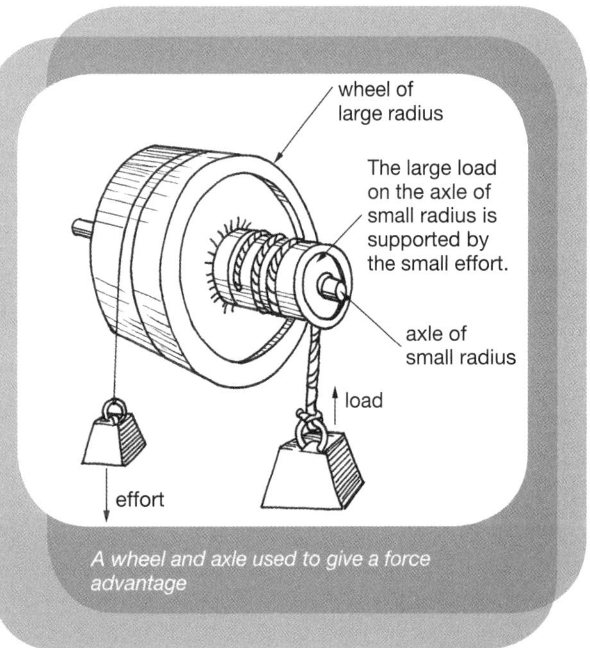

A wheel and axle used to give a force advantage

to act on the smaller wheel or axle. In this way a small force applied to the wheel can move a large load attached to the axle. This is because the effort distance is greater than the load distance since the wheel makes a bigger turn than the axle. This makes it easier to steer a car or truck and to tighten a screw using a screwdriver.

Inclined planes

An inclined plane like a ramp is a simple machine that allows heavy objects to be lifted or raised by smaller forces. When an inclined plane is used to raise an object the same amount of work is done by the smaller force moving a greater distance. Long planes with shallow slopes will lift things most easily.

Other examples of inclined planes are steps and stairs, switchback roads, screws, nuts and bolts and wedges like the head of an axe.

Pulleys

Pulleys are simple machines that can change the direction of a force and can give a force advantage. Pulleys usually consist of one or more wheels and a length of rope, although sometimes they use a chain or a belt. A single fixed pulley allows you to lift an object upwards by pulling down on a rope. You can easily make a bucket shower using a single fixed pulley. You can lift an object which weighs as much as you by using a single pulley. The branch of a tree can also be used as a single fixed pulley to lift something. The branch acts like a pulley wheel although there will be more friction than using a pulley wheel.

A single moveable pulley can be used to give a force advantage. You need less force to move the load with a pulley, but you need more rope. A single moveable pulley does not change the direction of a force. The load will move in the same direction as you pull but you will use a smaller force.

When we use pulleys we usually use fixed pulleys and moveable pulleys together as shown in the diagram below.

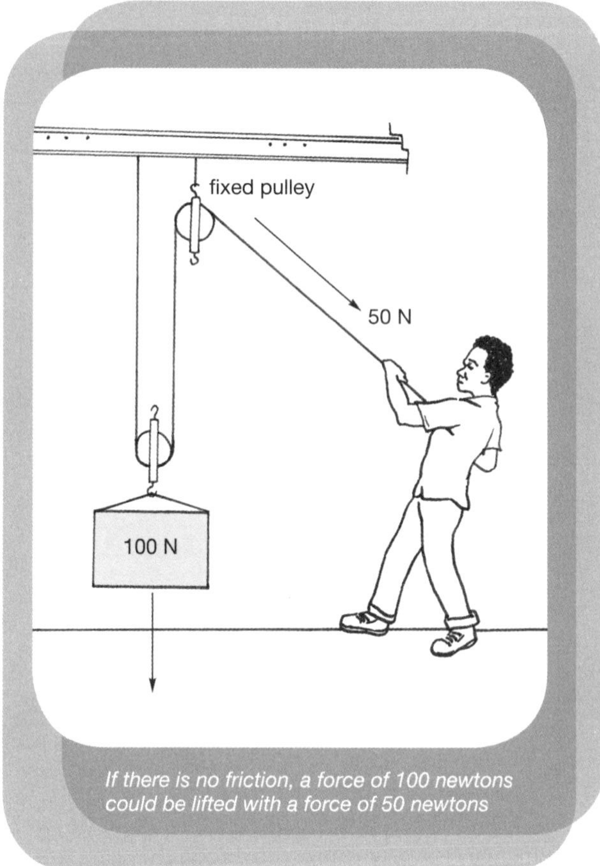

If there is no friction, a force of 100 newtons could be lifted with a force of 50 newtons

If there is no friction in the bearings of the pulleys, the combined single fixed and single moveable pulley will have a mechanical advantage of two. This can be worked out as follows:

$$\text{mechanical advantage} = \frac{\text{load}}{\text{effort}} = \frac{100}{50} = \frac{2}{1} = 2$$

Pulleys with more than one wheel are called pulley blocks. Pulley blocks can be arranged to give a large mechanical advantage. The approximate mechanical advantage of a set of pulleys can easily be found by counting the number of ropes supporting the load. For example, in the diagram above there are two ropes supporting the load so the mechanical advantage is two. The third rope comes from the top fixed pulley which is used to change the direction of the effort when we pull so we don't count this rope.

Pulley blocks allow you to lift a heavy load by using a small force but need a lot of rope, wire or chain to lift the object only a little way. Pulley blocks are used to control the sails on sailing boats, in cranes to lift heavy objects and in workshops to remove the engines from cars and trucks. A pulley block together with the rope or chain is also called a **block and tackle**. Sometimes the chain on a block and tackle is arranged in a continuous loop to make it easier to use.

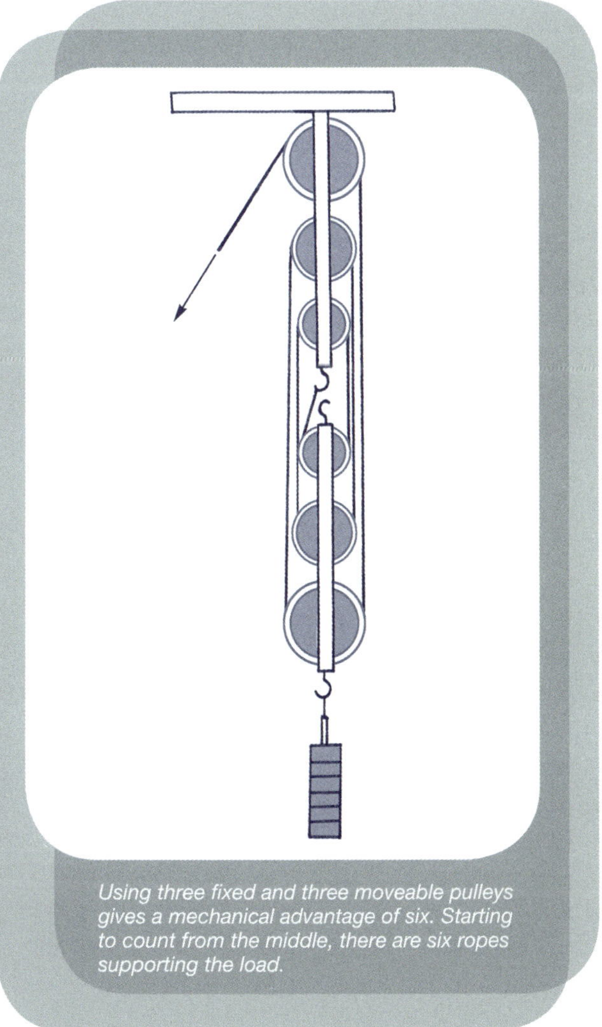

Using three fixed and three moveable pulleys gives a mechanical advantage of six. Starting to count from the middle, there are six ropes supporting the load.

For you to try

A

1 What is a simple machine? List five types of simple machines.

2 Copy the diagrams of the levers shown in the diagrams below. Mark on each diagram the pivot, effort, effort distance, load and load distance.

a) moving a log

b) moving a boulder

crowbar

small stone

c) elbow

(continued over)

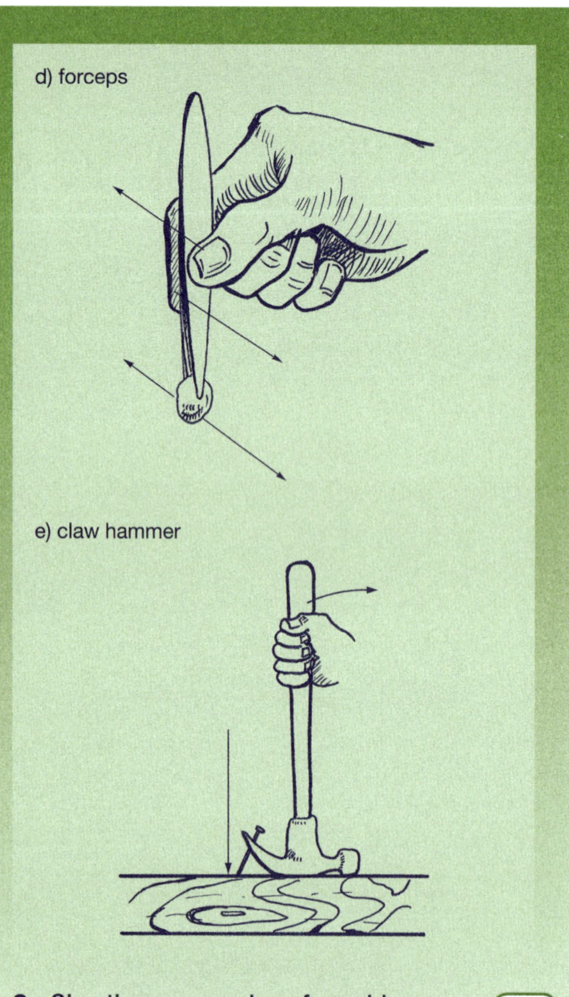

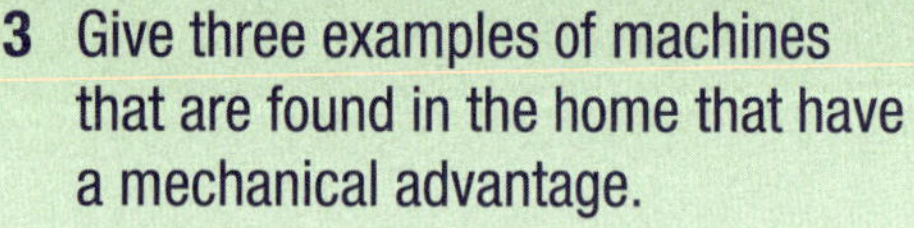

3 Give three examples of machines that are found in the home that have a mechanical advantage. 

4 If you use a force of 20 N to move a wheelbarrow 5 metres, how much work (in joules) do you do? 

5 A model crane lifts a mass of 4 kilograms through a height of 3 metres. How much work does the crane do? 

6 With your teacher, make a visit to a wharf, garage or building site to see how pulleys are used in a crane or block and tackle. 

Machines and efficiency 

When we use a simple machine we put in a force. The machine then puts out a force that does the job. Many simple machines give a force advantage. In other words they put out more force than you put in. However, the output movement is less than the input movement. We can see this more clearly when we try to move an object with a weight of 40 newtons using a lever. The output force is twice the input force, but only moves half the distance. So the work output equals the work input as shown below.

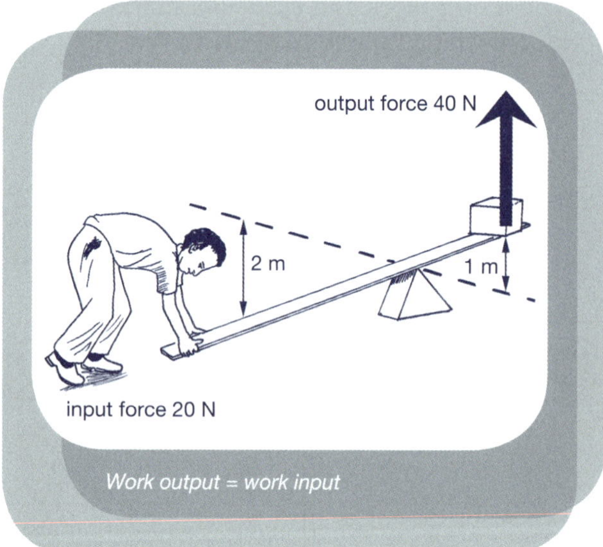

Work output = work input

work input	**work output**
= force x distance moved	**= force x distance moved**
= 20 × 2	**= 40 × 1**
= 40J	**= 40J**

When we compare the work input with the work output of a machine we can calculate the efficiency of the machine using the following equation:

$$\textbf{efficiency} = \frac{\textbf{energy output}}{\textbf{energy input}} \times 100\%$$

In the example above, the energy output is equal to the energy input. This means that the lever wastes no energy, so its efficiency is 100%.

Most machines waste energy because of friction in their moving parts. Some energy is also lost as heat. This means their efficiency is less than 100%. In other words the energy output is less than the energy input.

Some examples of efficiency values for machines and electrical appliances are shown in the table below:

For every 100J of input energy →	Output energy	Efficiency
Electric jug	99J	99%
Torch battery	90J	90%
Power station	35J	35%
Diesel engine	35J	35%
Petrol engine	25J	25%
Fluorescent tube or bulb	20J	20%
Human being	15J	15%
Filament light bulb	3J	3%

For you to try

1 Use some different examples to explain what we mean when we talk about the efficiency of a machine. What happens to the energy that is wasted? L

2 Explain why it is better to use fluorescent tubes or bulbs for lighting rather than filament light bulbs. L

3 Explain why it is important for engines and power stations to have the highest possible efficiency.

Summary questions

1 In the answer column in the table below write the letter of the meaning, A to K, which most closely matches with each physical property, I to XI.

Meaning		Physical property		Answer
A	Bends easily and can be hammered into shape	Brittle	I	
B	Can be pulled out into wires	Conductor	II	
C	Hard, but can break easily	Dense	III	
D	Lets heat or electricity pass through easily	Ductile	IV	
E	Lets some light through, but scatters it	Hard	V	
F	Not easily scratched or worn away	Lustre	VI	
G	Resists the effects of forces	Insulator	VII	
H	'See-through' or lets light through very easily	Malleable	VIII	
I	Shiny or lustrous	Transparent	IX	
J	Stops heat or electricity passing through	Translucent	X	
K	A lot of matter in a given space – dense objects sink in water	Strong	XI	

Questions 2 to 4 refer to the diagrams below which show containers of hydrochloric acid particles (A) and sodium hydroxide particles (B).

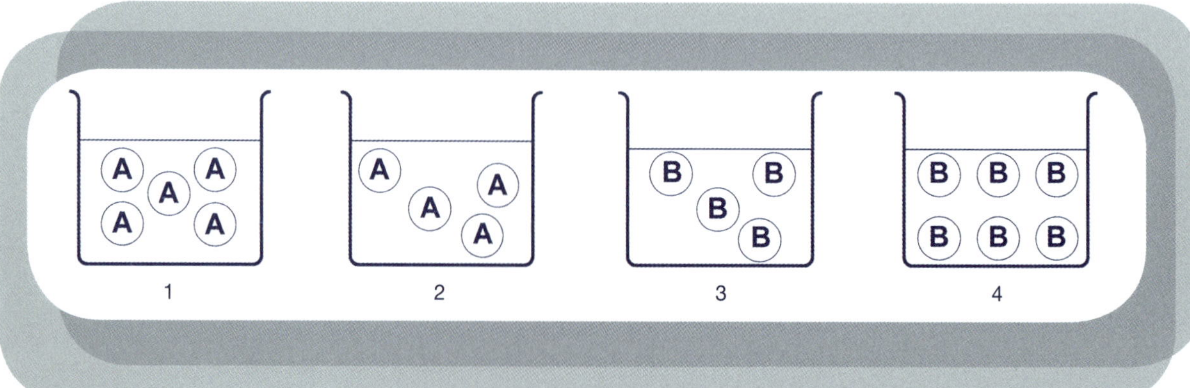

2 If two containers are mixed as follows when will an acidic solution be formed?

A 1 and 3

B 1 and 4

C 2 and 3

D 2 and 4

3 If two containers are mixed as follows when will a neutral solution be formed?

A 1 and 3

B 1 and 4

C 2 and 3

D 2 and 4

4 How many different basic solutions can be made mixing two or more of the containers?

A two

B three

C four

D five

5 Which of the following best describes the way that heat can move from a fire through the bottom of a saucepan?

A mainly conduction

B mainly convection

C conduction and convection

D conduction, convection and radiation

6 Which of the following best describes the way that heat can move in the water inside a kettle?

A mainly radiation

B mainly convection

C convection and radiation

D conduction, convection and radiation

7 Which of the following best describes the way that heat from the Sun reaches the Earth?

A conduction

B convection

C radiation

D conduction, convection and radiation

8 Which of the following are true?

I Some machines can put out more force than you put in.

II Some machines can put out more energy than you put in.

A I only

B II only

C I and II

D neither I or II

9 Put these forces in order from the smallest to the biggest where '1' is the smallest.

	Force	Rank
A	A truck hitting a pole	
B	An aeroplane taking off	
C	Threading cotton through the eye of a needle	
D	Kicking a ball	
E	Pushing a car along the road	

10 Which of the following are all effects of the force of gravity?

I holding things on the surface of the Moon

II making things fall back to the Earth

III holding satellites and the Moon in orbit around the Earth

A I and II only

B II and III only

C I and III only

D I, II and III

11 The Moon is much smaller than the Earth and has less mass. An object taken to the Moon weighs less than it does on the Earth. If the same object was taken to Jupiter, the largest planet, what would happen to its weight?

A it would weigh the same as it does on the Earth

B it would weigh the same as it does on the Moon

C it would weigh more than it does on the Earth

D it would be weightless because it is in space

Questions 12 to 14 refer to a rock that has a mass of 3 kilograms.

12 What is the most likely weight of the rock on the Earth?

A 0 N

B 1 N

C 5 N

D 30 N

E 60 N

13 What is the most likely weight of the rock on the Moon?

A 0 N

B 1 N

C 5 N

D 30 N

E 60 N

14 What is the most likely weight of the rock deep in space between the Earth and the Moon?

A 0 N

B 1 N

C 5 N

D 30 N

E 60 N

15 Which of the following best describes the main effects or uses of electricity?

A conductors and insulators

B series circuits and parallel circuits

C heating and lighting

D heating, lighting and magnetism

16 Which of the following are ways of producing electricity?

A dry cells, torches and radios

B generators, dynamos and electric motors

C lead accumulator batteries and hydro-electricity

D solar cells, light bulbs and fluorescent tubes

17 The passage below is a summary of the main ideas of this chapter. Copy and complete the passage in your book. Using the words in the list, find the words that are missing. You can use each word only once.

acids, bases, chemical, circuit, effects, energy, form, float, machines, materials, measure, mechanical, nature, properties

The materials that we use every day have different physical ______________________ such as hardness, strength and density. When we use ______________________ we choose those that have the most suitable properties. For example, some things will ______________________ and some things will sink. Some substances have different ______________________ properties. For example, some things are ______________________ and some things are bases. Acidic and basic substances can be found in ____________ and are used by people in different ways. Acids and ______________________ can also be made and used in everyday life.

When we use ______________________ in the home we often use heat energy and electrical energy. Heat is a ______________________ of energy that can make substances hot and can be controlled. Temperature is a ______________________ of how hot a substance is. Electricity has three main uses or ______________________—heating, lighting and magnetism. Electricity must flow around a path called a ______________________ which we can draw as a diagram. Simple machines can help people to do work because there is a ______________________ advantage. We can calculate the amount of work that we do and the mechanical advantage of different ______________________.

4

Earth and beyond

Chapter summary

In this chapter you will have an opportunity to:

- collect and describe stone tools and other rocks, and find out why some rocks are better for making stone tools than others
- find out more about the difference between sedimentary, igneous and metamorphic rocks
- show how igneous and metamorphic rocks are formed
- find out about space exploration and share this information with others
- make models of spaceships and explain their importance in space travel and communication
- make a simple telescope and explain how it works.

Syllabus references

Strand: Earth and beyond

Sub-strand: Our Earth and its origin
Space exploration

Outcomes:

8.4.1 Demonstrate the formation of igneous and metamorphic rocks and relate findings about the properties of rocks to the ways they are used

8.4.2 Collect information about the human exploration of space

Key facts

- All rocks are made of minerals. The way a rock looks and its properties depend on the minerals that make the rock.
- Rock can be formed in different ways—igneous, sedimentary and metamorphic.
- Igneous rocks are formed by the cooling of molten rock called magma.
- Slow cooling of magma produces large crystals, while fast cooling produces small crystals.
- Igneous rocks can be classified according to crystal size, minerals and colour.
- Sedimentary rocks are formed from broken-down rock that is eroded and carried by rivers to the sea where it collects and becomes cemented together again.
- Metamorphic rocks are rocks that have been changed.
- Metamorphic rocks can be formed by heat and pressure. The type of metamorphic rock produced depends on the amount of heat, the amount of pressure and the type of original rock.
- Rocks can be classified according to the minerals they contain, the size of the crystals and the way that we use them.
- Metals can be obtained from minerals and the extraction of metals from rocks known as ores is important in Papua New Guinea.
- We use satellites orbiting the Earth in space to transmit radio, television and telephone messages to any part of the world. Satellites are important in communication.
- We can find out more about space by using different kinds of telescopes.
- People have travelled through space to the Moon and there is an international **space station** orbiting the Earth which a few people from different countries visit for short periods of time.

Rocks and their importance (PD) (SS) (A)

Different kinds of rocks and metals obtained from rocks have been important to people in many parts of the world for thousands of years. When we try to find out about the way that people lived in the past we can see that many of the things they made came from materials they found in the ground. Scientists think that people in different parts of the world developed through three stages or ages, learning first to make tools from stone and then metals, as shown in the table below.

The three ages did not happen in every part of the world and did not happen at the same time in different places. However, scientists have enough information from different parts of the world to understand the way in which people slowly learned to make metals from rocks.

Stones that could be sharpened were used to make axes for cutting down trees (on the right) and adzes for hollowing out canoes (on the left).

Age	How many years ago?	What materials did they use?	What did they make?	What else did they do?
Stone age	about 2000 to 3500 BC	stone, clay, wood	axe, scraper, spear, bow and arrow, boats, pottery in some places	hunting and gathering, fishing, start of agriculture, living in caves and huts near water
Bronze age	about 3500 to 500 BC	metal ores to make copper and tin and mixing them to make **bronze**	axes, knives, tools for agriculture, ornaments for decoration	agriculture, cattle breeding, potter's wheel to make pottery, trade
Iron age	about 500 BC to 500 AD	metal ore to make iron and steel	iron tools, spears and swords, tools for agriculture	first cities begin

Rocks which do not break when heated in fire can be used for making a mumu.

Minerals

Rocks and everything else in the world are made of chemical elements. Gold, silver and copper are examples of elements that can be found as metals in nature. However, not many elements are found in a natural state in the environment. This means that usually we cannot just dig them out of the ground. Most substances in the Earth's crust are formed from elements that are joined together. These substances are called chemical compounds.

The chemical elements and chemical compounds that make up rocks are called minerals. There are many different minerals but only a few are common in the Earth's crust.

All rocks are made of minerals. A rock's appearance and properties depend on the minerals that make up the rock. Most rocks are a mixture of a few minerals. For example, the **igneous rock** called **granite** contains the minerals quartz, feldspar and mica. Some rocks contain minerals that are very useful to the community. For example, the minerals may contain metals such as iron, gold and copper. Rocks that contain metals are called **mineral ores**. One of these is chalcopyrite which is a copper ore that is common in Papua New Guinea and is usually yellow in colour. The table below shows some mineral ores and the metals obtained from them.

Some mineral ores and their related metals

Mineral ores	Metal obtained
haematite	iron
limonite	iron
bauxite	aluminium
chalcopyrite	copper
azurite	copper
galena	lead
cinnabar	mercury
cassiterite	tin
carnonite	uranium

Igneous rocks

Deep inside the Earth the temperature is very high and the rocks may melt. These **molten** rocks are called **magma**. When magma cools and solidifies the rock that is formed is called igneous rock. In most cases the magma goes solid before it reaches the surface. Igneous rocks that are formed inside the Earth are called **intrusive rocks**. Sometimes the magma reaches the surface of the Earth through a volcano (the molten rock is then known as lava) or flows through a deep crack called a **fissure**. Igneous rocks formed in this way are called **extrusive rocks**.

An igneous intrusion. The dark-coloured intrusion has been pushed up into the lighter-coloured rock.

Igneous rocks found in different parts of the world come from different places inside the Earth and so contain different minerals. This means that igneous rocks can be classified by the type of minerals they contain and by the size of the crystals of the minerals.

Crystal size

When magma cools quickly it forms crystals which are very small. If the cooling is very fast, then an igneous rock called **obsidian** is formed. Obsidian looks like black glass and can be made into tools with a sharp edge. Obsidian from Lou Island in Manus Province has been found in many parts of the south west Pacific showing that it was traded and then used to make spears, scrapers and other sharp tools.

The igneous rock called **basalt** is formed when lava cools quickly on the surface of the Earth. Basalt rocks can be seen near Sogeri, Lae, Mount Hagen and in many other places in Papua New Guinea.

A special igneous rock called pumice is sometimes found washed up on beaches. Pumice is formed when gases dissolved in the rock are released when the lava erupts at the surface through a volcano.

Obsidian

Pumice floating on water

The lava hardens very quickly and bubbles of gas form many small holes in the rock which make the pumice light enough to float on water. When the volcanoes in Rabaul erupted in September 1994, Simpson Harbour was covered in pumice. Pumice stone can also be used to clean things and to rub on your skin to make it soft or clean it.

Magma below the surface of the Earth cools slowly and forms large crystals that can easily be seen with the naked eye. For example, granite is an igneous rock that has formed by slow cooling deep in the Earth. It may take thousands of years for a large pocket of magma to cool and solidify.

The continents are mostly formed from light-coloured rocks such as granite, but the sea-floor is formed from dark-coloured rocks such as basalt. Because granite is hard with closely packed crystals it can be polished easily. In some countries it is used in building and for making monuments and headstones for graves.

Granite

Sometimes within the Earth pockets of magma are formed that may be under great pressure. This pressure is able to force the magma into cracks in other rocks close to the surface. This will cause the magma to cool more quickly, giving rocks with smaller crystals. The igneous rock called dolerite is formed in this way and is sometimes used to make roads because it is hard.

Different names are given to rocks that have the same composition but different crystal sizes. For example, granite and obsidian contain the same minerals, but because they cooled at different rates they have different crystal sizes and so are given different names.

Classifying igneous rocks

Igneous rock may vary in colour from almost white to black. The colour indicates the minerals the rock may contain. Light-coloured rocks such as granite and pumice contain light-coloured minerals such as quartz. Dark-coloured igneous rocks contain dark-coloured minerals which are usually heavy. The minerals that are found in a rock are described as the mineral composition, which is important in helping to classify rocks. Igneous rocks can be classified according to the minerals they contain and their crystal size:

- Igneous rocks with small crystals have cooled quickly and are said to be fine-grained.
- Igneous rocks with large crystals have cooled slowly and are said to be course-grained.

The table below can be used to classify igneous rocks.

Classification of igneous rocks

Crystal size	Light-coloured → dark-coloured		
Small crystals Fine grain	rhyolite	andesite	basalt
Large crystals Coarse grain	granite	diorite	gabbro

For you to try

1 **Investigation: Using stones**
   a Collect some stone tools such as an axe, adze, a scraper and some mumu stones.
   b Explain how each stone is used. What are its useful properties?
   c Examine the characteristics of the stone and try to identify the rock from which it is made.
   d Make some conclusions about why these rocks were used and not other rocks.
   e Make a drawing of each tool and add some labels that explain how and where the tool was used.

2 Explain how igneous rocks are formed.

3 What is the difference between magma and lava?

4 Answer the following questions about pumice.
   a What is pumice?
   b Where is it likely to be found?
   c How does it form?
   d What can it be used for?

5 How does the cooling of igneous rock affect the size of the crystals and the grain of the rock?

6 Why are igneous rocks sometimes used to make buildings, for gravestones and as material for making roads?

7 What characteristics are used to classify igneous rocks?

Sedimentary rocks

Sedimentary rocks are formed from layers of **sediment** deposited by the sea, rivers, wind or moving ice. The sediments are squashed as more and more material collects above them. Then they become hard, which is similar to the way in which concrete becomes hard. This is a slow process and can take millions of years. The layers of rock are called strata. You can see layers in sedimentary rock such as limestone, sandstone and **coal**.

Most sediments are particles of eroded rock. Sandstone is formed from sediments like this. However, some sediments are fragments of shells and bones from animals that lived hundreds of millions of years ago. Limestone is usually formed in this way, although it can be deposited chemically like the scale in a jug or kettle. Limestone can be used in buildings, to make roads and in making cement.

Metamorphic rocks

Many things change their form. For example, a tadpole changes into a frog, and a caterpillar changes into a butterfly. This process is called metamorphosis. Rocks can also change when they are affected by heat and pressure. For this reason these rocks are called **metamorphic rocks**.

Metamorphic rocks have been formed deep underground from sedimentary rocks and igneous rocks by the action of heat or pressure, or both. The metamorphic or changed rock is usually harder than the original.

How metamorphic rocks are formed

Pressure

Movement in the Earth's crust creates great pressure within the rocks of the crust and this can make rocks change. The minerals may be flattened

The Earth's rocks

and pushed into layers. Slate is a metamorphic rock produced by pressure from shale or mudstone. Its layers are so well formed that it can be split into sheets. These split sheets can easily be cut to size and have been used for hundreds of years as roof tiles and floors of houses in some countries. Slate roof tiles are not heavy and can be drilled and nailed on to the wooden frame of the roof. Slate is also used to make pool tables because a big sheet of slate is flat and smooth.

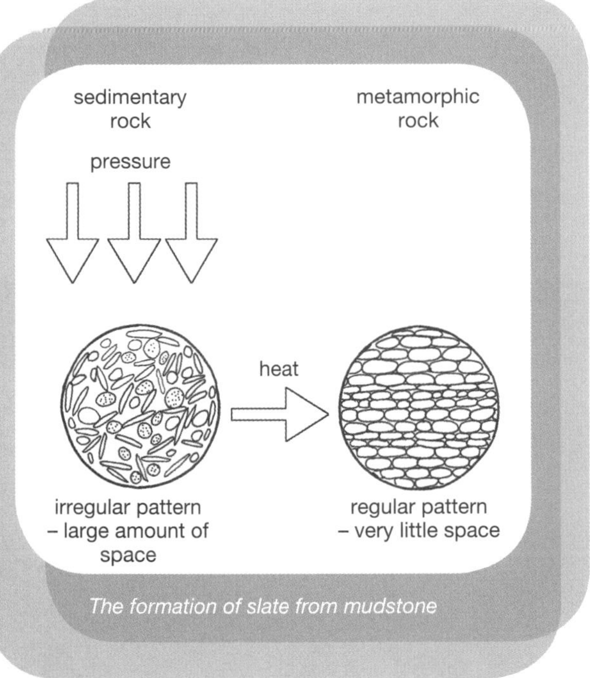

The formation of slate from mudstone

Slate can be split and cut to make roof tiles and floor tiles.

Heat

When rocks are heated the materials within the rock may be changed. The way in which traditional cooking pots are made is similar to the way in which some metamorphic rocks are formed. Clay is heated to a high temperature and the materials in it are changed so that the clay becomes hard and strong.

Rocks may be heated when magma comes into contact with them. The amount of change depends on the temperature and the distance of the rock from the magma. Rocks that are heated to very high temperatures and closest to the magma will change the most. This type of metamorphosis is called contact metamorphism and most contact metamorphic rocks are called **hornfels**.

The sedimentary rock limestone can be changed into **marble** by metamorphosis. In this case new crystals are formed. People living in countries around the Mediterranean Sea such as Greece have been using marble for hundreds of years to make buildings and statues. Marble is also mined in North Queensland and Western Australia. Marble can be cut and polished and lasts a long time but is very expensive. Sometimes **fossils** can be seen in marble that has been polished. Marble has been used on the floor of the National Parliament building in Port Moresby and is also used in bathrooms and kitchens.

Marble is used to make statues.

Marble floors in the National Parliament, Port Moresby

For you to try

1. Describe the different ways in which metamorphic rocks can be formed. (L)
2. Describe the process involved when slate is formed. Why does slate split so easily? How is slate used? (L)

3 Explain how marble is formed and how it can be used. Why does marble sometimes contain fossils? (L)

4 Slate is sometimes used to make roof tiles. Describe the advantages and disadvantages of making a roof using 
   a granite
   b limestone

5 If you live near one of the war cemeteries or another old cemetery arrange a visit with your teacher to look at the gravestones and other memorials. Make a drawing of some of the gravestones or memorials. Try to identify the type of rock used and the age of the gravestones or memorials. (SS)

Metals from minerals

Metals are found in the minerals that occur naturally in rock. Papua New Guinea has many valuable mineral deposits, especially of copper, silver and gold. The table below shows some of the minerals found in Papua New Guinea and the metals that are extracted from them.

Minerals found in PNG and the metals extracted from them

Mineral	Formula	Metal	Formula
niccolite	$NiAS$	nickel	Ni
linnaerite	Co_3S_4	cobalt	Co
pyrolusite	MnO_2	manganese	Mn
argentite	Ag_2S	silver	Ag
molybdenite	MoS_2	molybdenum	Mo
gibbsite	$Al(OH)_3$	aluminium	Al
silver	Ag	silver	Ag
gold	Au	gold	Au
chalcopyrite	$CuFeS_2$	copper	Cu

Metal ores

The rocks from which we obtain metals are called **ores** and a large amount is known as an **ore body**.

In the countries around the Pacific Ocean there are many volcanoes and earthquakes and this is sometimes known as the rim of fire. The changes occurring inside the Earth that cause earthquakes and volcanoes also bring rocks containing metal ores close to the surface. Some rocks contain large enough amounts of one metal compound or one metal so it is worth digging up and extracting the metal. For example, gold has been mined in the Pacific since 400 AD when Chinese miners worked in Indonesia. Gold was found in Fiji in 1868 and was observed in local pottery in Redscar Bay, Central Province, in 1852. By the 1890s there were hundreds of gold miners in the islands of Milne Bay and the money earned from gold sold overseas was very important. We now know that there are thirty large mineral deposits of gold in the Pacific and twelve of these are being mined or are expected to become mine sites. The map on page 116 shows mines and possible future mines in Papua New Guinea.

Searching for metals

Scientists use a number of different methods in order to explore and find ore bodies:

- Aerial surveys—looking at geological maps to find rock types and taking photographs from aeroplanes. Helicopters can also be used to carry equipment to measure the magnetism

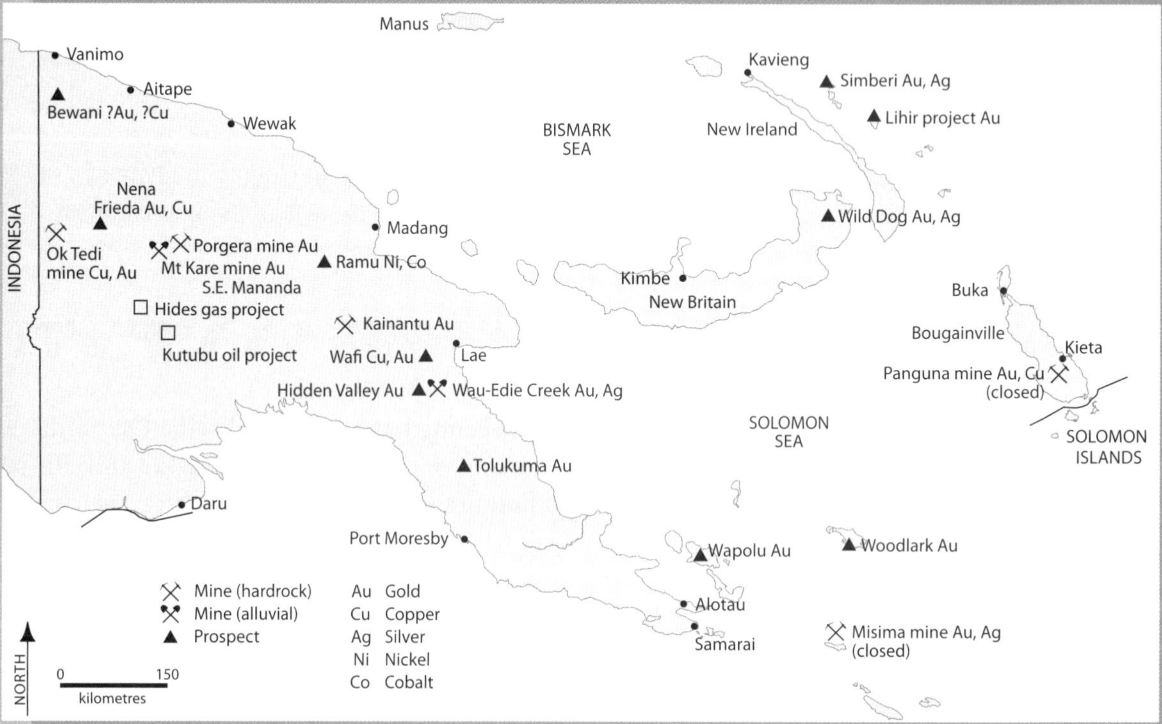

Mines and prospects in Papua New Guinea

and conductivity of the ore body as they fly over.

- Surface exploration—small amounts of metal ores will dissolve in water and can be found in rivers. By testing the water at different places scientists can find the ore body, as shown in the diagram on page 117. The copper ore bodies at Panguna and Ok Tedi were found in this way.

After scientists find an area that they think may contain an ore body they take samples of rock and carry out tests to identify the minerals and the amount of metal present in the ore. This will help decide whether to build a mine and take out the ore body.

A helicopter conducting a survey to measure the conductivity

Scientists would find small amounts of zinc in water collected from D, E and G, but not from A, B, C, or F. This would help them to find the ore body.

Extracting metals

Most metals occur as compounds mixed with other minerals in rocks. The main stages in obtaining metal from their ores are usually mining, concentration, smelting and refining.

A few metals are found naturally occurring in the environment. For example, gold is usually found in veins in rocks containing quartz. When the rock is weathered and eroded the sand and gravel formed is mixed with small pieces of gold and carried by rivers where it settles on the bottom. This is called **alluvial gold**. Gold found in this way is known as a **placer deposit** and is usually the first to be discovered. We can separate the gold by a process called **panning** which is described in the box below. There are about 4000 small-scale miners in Papua New Guinea earning their living by panning for gold.

Science in the village: Panning for gold

Panning is a simple way to obtain gold that has been used for over one hundred years. In 1989 there was a gold rush at Mt Kare and many Papua New Guineans including children stopped their normal activities to look for gold in creeks in the local area. The only piece of equipment they needed was a metal pan, which is big dish with a flat bottom and sloping sides that looks a bit like a big frying pan without a handle.

First of all the people had to know where to look. Gold is much heavier than water and stays on the bottom and gets caught in the sand in slow moving areas around bends and along the bank of the creek or river. Gold also gets stuck in the cracks in rocks and in pieces of wood, so these are good places to look. It is also a good idea to go to a creek or river where other people have found gold before.

After choosing a good place to look, these are the steps they followed in panning for gold:

- Put about four handfuls of sand and gravel into the gold pan and put the pan under water.

(continued over)

- While holding the pan under water, move the pan in a circular motion so that the lighter materials are carried out of the pan.
- It is important not to move it too quickly or you may lose the gold.
- Keep doing this until about half the material in the pan is gone.
- Lift the pan out of the water and move it round in a circle with one edge tipped slightly down.
- When all the water is gone dip the pan in the water again, lift it out and start swirling again.
- Keep doing this until all sand and gravel is gone.
- Small pieces of gold will be found in the bottom of the pan if you are lucky.
- The small pieces of gold are then carefully picked up and placed in a small container.

Panning for gold takes lots of practice and lots of time so you have to be patient. You might also need to be lucky but many people at Mt Kare earned a lot of money from the gold they found.

Panning for gold in Papua New Guinea

The importance of minerals in Papua New Guinea SS

In Papua New Guinea the people own the land but they do not own the minerals found 20 metres or more below the surface. It is the national government or state that owns all the minerals found below this level. For this reason the Department of Mining and Petroleum is able to control all mineral exploration in the country. This department gives out licences to companies to look for minerals. These companies often spend millions of kina searching for minerals and sometimes this may lead to a mine being started.

The government is able to earn a lot of money from the goods that are sold overseas. In 2005 about half the money that the government earned came from the mining of minerals. For this reason, the mining of minerals has become very important to the people of Papua New Guinea. However, minerals are a non-renewable resource and each mine has only a fixed life time. For example, copper is produced only at Ok Tedi since the mine at Panguna in the North Solomons closed in 1989. Gold was produced at Misima in Milne Bay until 2004. A new gold mine at Kainantu began in 2005 and exploration continues in about eighty other places, with drilling taking place in about ten of these.

Mine	Province	Copper (t)	Gold (kg)	Silver (kg)
Ok Tedi	Western	194 355	17 110	45 898
Porgera	Enga	–	16 279	3242
Lihir	New Ireland	–	20 249	–
Tolukuma	Central	–	1024	1958
Kainantu	Eastern Highlands	–	687	n/a
Small scale	various	–	3000	n/a
Total		194 355	58 349	51 098
Export Value (K million)		4329.5	3090.9	25.2

For you to try

1 List some of the metal objects that are found in your home or community.

2 What properties of metals make them so useful?

3 What are the advantages and disadvantages of a roof that is made from bush materials and one that is made from corrugated iron?

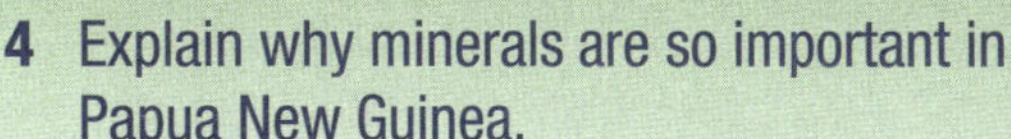

4 Explain why minerals are so important in Papua New Guinea.

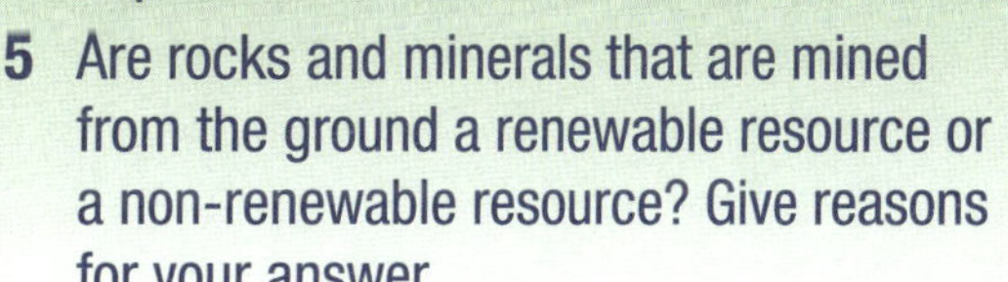

5 Are rocks and minerals that are mined from the ground a renewable resource or a non-renewable resource? Give reasons for your answer.

The table above shows the main mines in Papua New Guinea and the amount of copper, gold and silver produced in 2006.

The people who live on the land where the mines are situated are also paid some money called **royalties** but mines can also cause problems like pollution.

The map below shows the existing and possible mines in Papua New Guinea.

Existing and possible mines in Papua New Guinea

Science in the village: Making lead weights

ML SS

Making metals from metal ores is difficult and needs a supply of a suitable metal ore, a very hot fire as well as special equipment, knowledge and skills. We don't know if people in Papua New Guinea learned to make their own metals in this way in the past, but it is unlikely. However, people have learned to work with some metals by heating them until they melt to make castings. For example people on the coast make their own fishing weights from old lead batteries using the following steps:

Molten lead being poured into a betel nut mould

- Collect old car and truck batteries and carefully pour away the acid making sure that it doesn't burn anyone or anything. Rinse with plenty of water to get rid of the acid.
- Break open the case of the batteries and remove the lead plates.
- Break up the plates and put them into an old pot or saucepan.
- Put the saucepan on a wood fire and heat strongly.
- When the lead melts it is poured into a suitable mould like the skin of a betel nut.
- People usually put a piece of bent wire inside the mould so that the fishing line can be tied on to the weight.

For you to try

Project: Panning for gold

Collect a large old frying pan and a small plastic container with a lid.

Visit a local creek or river and try panning for gold.

Use the information in the box on pages 117–118 to help you choose a good place to look and to find out how to use the pan.

Project: Making fishing weights

Collect old batteries and try making your own fishing weights using the steps in the box above.

Space exploration (SS)

Different kinds of rocks and metal ores obtained from rocks on the Earth have been important to people in many parts of the world for thousands of years. In the last fifty years people have also begun to explore beyond the Earth and to send rockets into space. One way of doing this is to send satellites into space, which then travel in orbit around the Earth. These artificial satellites of the Earth behave in the same way as the Moon, which is a natural satellite of the Earth. The further away they are, the longer they take to make one complete orbit.

The first artificial satellite was sent into orbit around the Earth in 1957 and since then thousands of satellites have been launched. There are about 560 working satellites orbiting the Earth but there are also thousands of old satellites and pieces from old rockets that orbit the Earth. If they fall back into the atmosphere they burn up due to the friction with the air.

Satellites of the Earth

There are two main types of satellites: **geostationary satellites** and other satellites. Geostationary satellites are put into orbit above the equator of the Earth at a height of about 36 000 kilometres. At this height they make one complete orbit of the Earth every twenty-four hours. The Earth also makes one complete rotation on its axis every twenty-four hours and so these satellites seem to remain in the same position above one point on the equator. This is useful because we can use a satellite dish to pick up a signal from the satellite without having to move the dish. Satellite dishes in Papua New Guinea that pick up television signals work in this way.

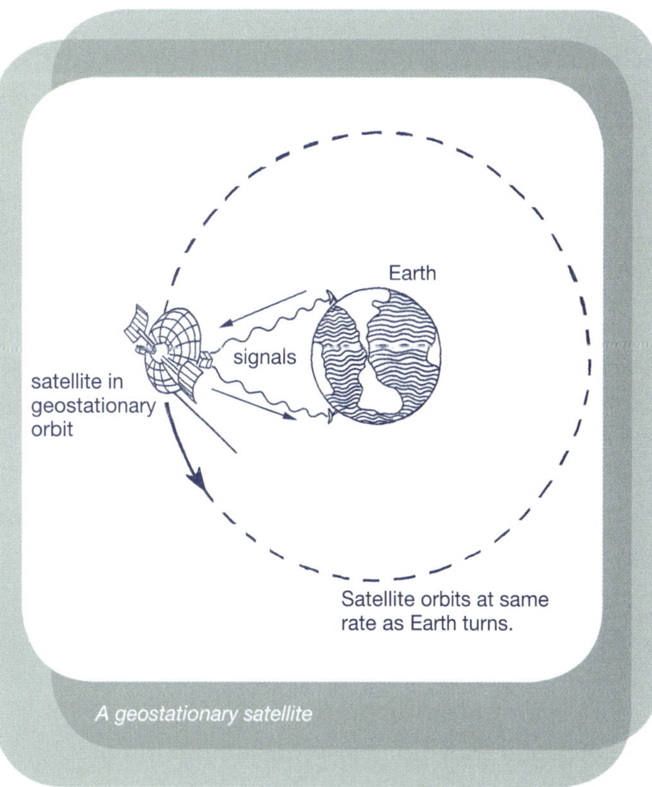

A geostationary satellite

One geostationary satellite can cover about one third of the Earth's surface. This means that only three satellites are needed to cover the whole Earth, as shown in the diagram on page 122.

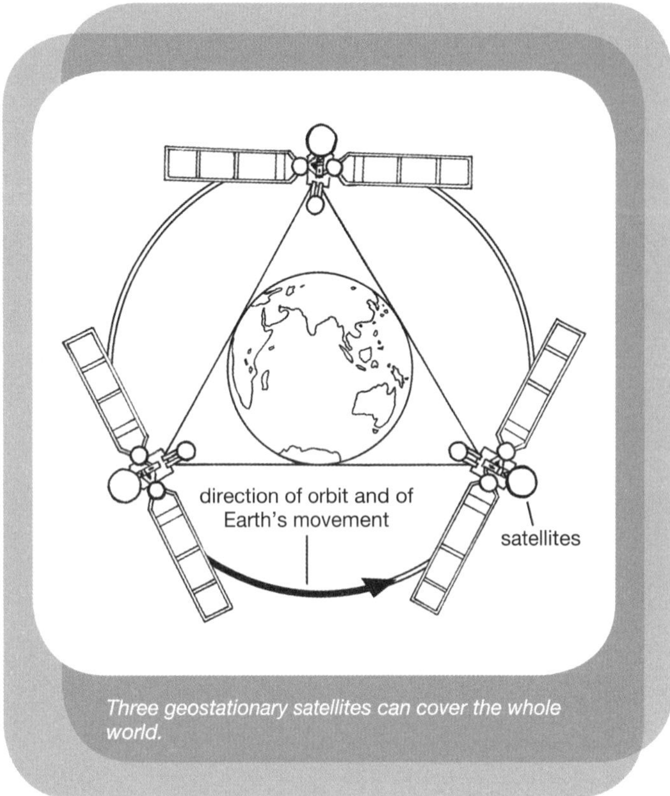

Three geostationary satellites can cover the whole world.

There are many satellites closer to the Earth that orbit in less than twenty-four hours and are not geo-stationary. The closest ones are about 160 kilometres above the Earth. There are also many satellites at about 1000 kilometres above the Earth. If you look at the sky on a clear night you can often see satellites moving steadily across the sky.

Uses of satellites

Satellites are used for a variety of different reasons.

- **Communication** satellites are used for telephone, radio, television and the Internet. For example, people in Papua New Guinea can talk with people in other countries and watch television programs from other countries.
- Navigation satellites use radio signals so that people on the ground, in a boat or an aeroplane, can find out exactly where they are and how to get to another place. This is called the Global Positioning System, or GPS, and is accurate to within a few metres.
- Mapping satellites help to make maps. The education department uses satellite maps to find the position of all the schools in the country.
- Weather satellites are used to watch and forecast the weather in different parts of the world.
- Astronomical satellites are used for observing planets, galaxies and other objects in outer space.
- Spy satellites and defence satellites are used by some countries to protect themselves against other countries.
- Space stations are satellites that are designed for people to live in space. Space stations are not usually able to move themselves or to land, so people must travel to them in another space craft.

Unmanned exploration of space

Unmanned lunar, solar and interplanetary satellites and probes have been sent to the Moon and into the **solar system** to gather information and send back photographs. A **space probe** is a spacecraft that flies by or lands on the surface of a body in the solar system. A body in the solar system is a planet or a natural satellite of a planet, which we call a moon.

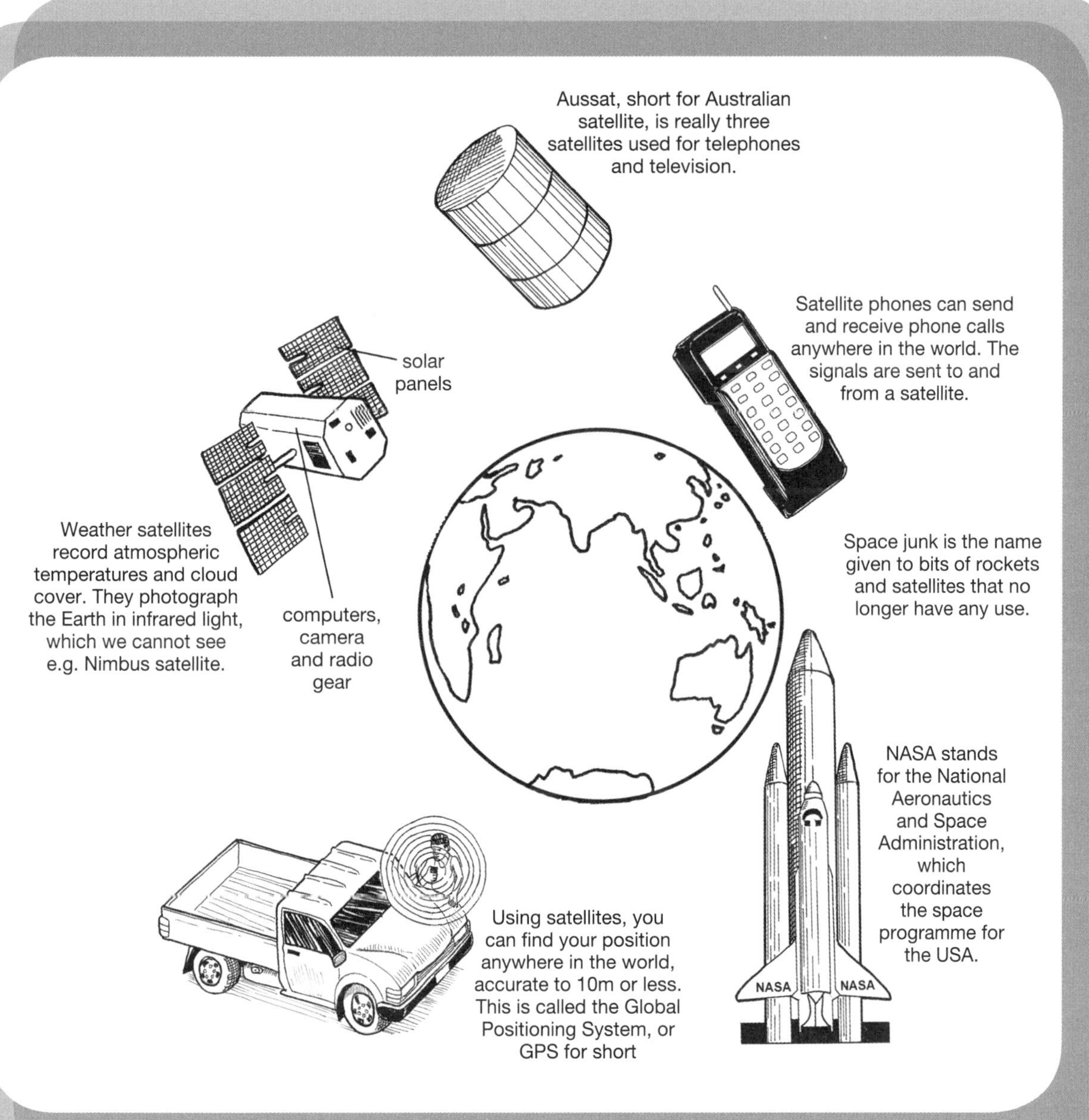

Some ways that satellites are used

Some unmanned space probes are shown in the table below:

Name of probe	Date	Purpose
Mariner (USA)	1962–73	Fly by Mercury, Venus and Mars
Surveyor (USA)	1966-68	Landed softly on the Moon, took pictures and found out about the surface
Venera 7–14 (USSR)	1970–81	Several probes to fly by Venus and land on the surface. Venera 7 was the first space craft to land successfully on another planet and send back information.
Pioneer 10 and 11 (USA)	1972–3	Fly by Jupiter and Saturn and beyond the solar system; each carried a small metal plate with a picture of a man and woman, and other information about the origin of the space craft in case it was found by other intelligent living things.
Viking (USA)	1974	Two probes that orbited and landed on Mars
Voyager 1 and 2 (USA)	1977	Fly by Jupiter, Saturn, Uranus and Neptune
Vega 1 and 2 (USSR)	1984–86	Landed on Venus and also collected information about Halley's Comet in 1986
Phoenix (USA)	2008	To land on Mars and explore below the surface
New Horizons (USA)	2006–15	Fly by Jupiter in 2007, fly by Pluto in 2015, then leave the solar system

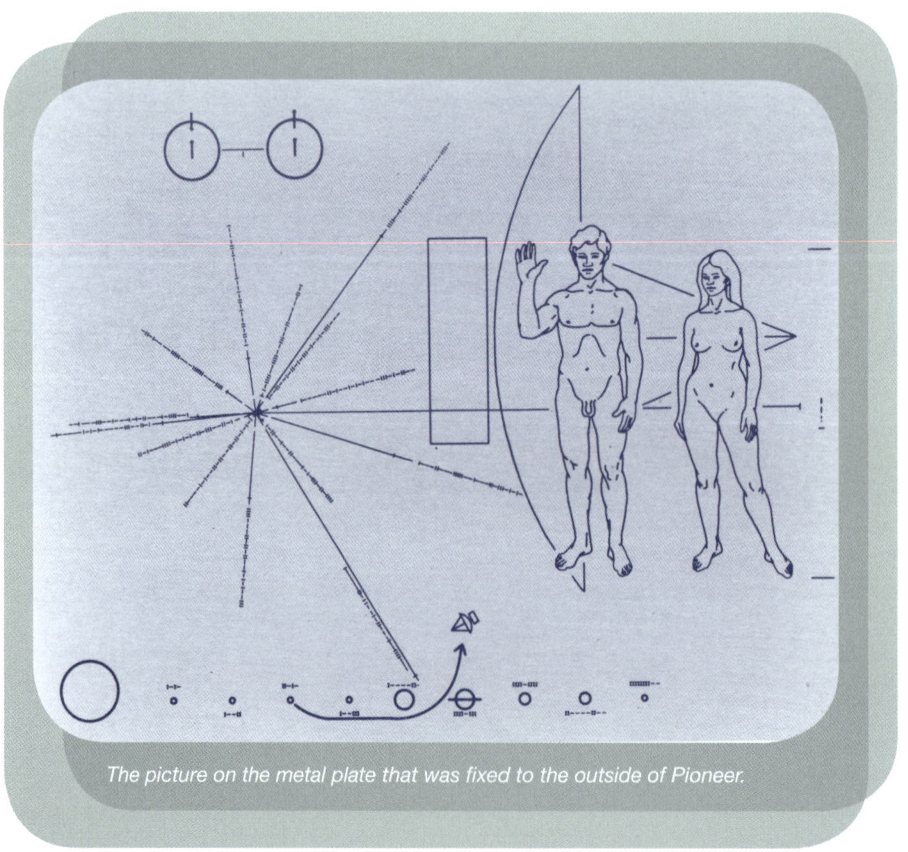

The picture on the metal plate that was fixed to the outside of Pioneer.

Humans in space

The first person in space was the Russian Yuri Gagarin, who made one orbit of the Earth in April 1961. The first American in space was Alan Shepard in May 1961 and John Glenn was the first American to orbit the Earth in February 1962.

The early manned spacecraft from America and Russia were small, closely-packed capsules that carried one or two people into orbit for a short time. In all of these programs big rockets were used to carry small spacecraft away from the Earth. The rockets and the spacecraft were not able to be used again.

The American Apollo project was the first to send people to the Moon. The first astronauts landed on the Moon in 1969 and twelve men had walked on the Moon's surface by the end of this project in 1972.

By the middle of 2007 there had been 258 manned spaceflights, of which 250 reached orbit successfully. Hundreds of people have travelled in space and about forty of these have been women. A few people who are very rich have paid millions of Kina to be a tourist in space.

The space shuttle

The American **space shuttle** is a space transport system designed to go into orbit around the Earth. The shuttle is launched into orbit by its own giant engines and two booster rockets. The shuttle can remain in orbit around the Earth for up to thirty days and return to the Earth and land like an aeroplane. The shuttle is the first spacecraft designed with parts that can be re-used. Both the shuttle and the booster rockets can be used up to 100 times, which helps to reduce the very high cost of space travel. The shuttle normally carries a crew of between five

Astronauts walking on the Moon. The Apollo landing craft is in the background and the wheels of the Moon rover can be seen on the right.

and seven people but has carried eight and could carry eleven people in an emergency.

Five space shuttles have been built and in 2007 there were three remaining. The first space shuttle was called Enterprise. It was built in 1977 but was used only for testing on the back of a jumbo jet and not for actual space flight. The next four space shuttles were called Columbia, Challenger, Discovery and Atlantis. Challenger was destroyed when it was being launched in 1986 and all seven crew members died. Endeavour was built as a replacement for Challenger. Columbia was destroyed when it was re-entering the Earth's atmosphere in 2003. Again all seven crew members died.

Much of the equipment and technology used on the space shuttle is over thirty years old and out of date so the space shuttle will not be used after 2010. It will be replaced by Orion which is a new spacecraft that will take humans to the Moon and beyond.

The American space shuttle *Atlantis*

Uses of the space shuttle

The space shuttle has been used in many ways:

- carrying crew members to and from the Mir space station and the International Space Station.
- carrying cargo to and from the International Space Station.
- allowing astronauts to repair the Hubble Space **Telescope**.
- carrying out experiments while in orbit around the Earth—especially since there is no gravity and people and things are weightless.
- carrying satellites into low orbit around the Earth.
- carrying parts of the International Space Station to make it bigger.
- carrying satellites to a point where they can then use their own engines to go into higher orbit around the Earth or into orbit around another planet—these include an observatory and defence and communication satellites.

The first space stations

Russia and America have both had space stations where people have lived for short periods of time:

- Russia had seven space stations called *Salyut* between 1971 and 1986.

- America had the huge space station called Skylab in 1973 and 1974 which weighed 77 tonnes. In 1979 it re-entered the Earth's atmosphere and burnt into small pieces which fell in Western Australia.
- The space station called Mir was controlled by the USSR from 1986 to 1991 and then by Russia until 1999. Different people lived on Mir for nearly ten years.

International Space Station

- The International Space Station is the largest and most difficult science project in history. Those taking part come from America, Russia, Canada, Japan and Europe.
- It is being built in space by sending the different pieces of it into space, which are then put together or assembled by astronauts.
- The project began in 1998, the station is expected to be completed in 2010 and be used until about 2016.
- When complete it will be about 110 metres across and about 90 metres long and weigh about 500 tonnes. The area of solar panels will be about the same size as a football field and it will have six laboratories for doing experiments.
- It orbits the Earth about sixteen times each day at a height of about 330 km and can be seen with the naked eye.

The International Space Station

- The first people began living in the space station in 2000 and a few people take it in turns to live there. This means that there is always a small group of people living in the International Space Station.
- Up until the middle of 2007, astronauts from fourteen countries had visited the space station, and also five space tourists.
- People living on the space station will carry out experiments to find out more about
  - the effect of weightlessness on the human body
  - the technology needed to live in space
  - exploring other planets and living on other planets
  - new ways to treat diseases
  - better ways to make different materials
  - measuring things more accurately than we can on Earth.

Living in space

In order to live and work in space, humans must deal with some interesting problems which we do not find on the Earth.

- Weightlessness: Because there is no gravity everything floats about unless it is fixed. Weightlessness also affects the way the human body works.
- Water: All human beings need water to live and astronauts need extra water when they live in space. They carry some water but, because it is very heavy, they also make it from oxygen and hydrogen. Urine is also collected and made pure so the water can be used again.
- Food: This must be healthy, nutritious, light in weight and must not break easily into small pieces that can float about. Dried food is often used and water is added to the food before it is eaten.
- Toilets: Normal toilets cannot be used because everything would just float about. Special vacuum toilets have arms which hold the person on to the seat and then the waste is sucked into a plastic bag.
- Washing: Astronauts cannot wash in the normal way because the water would just float about and get in their eyes and even up their nose. They have a sponge bath using a damp sponge or cloth.
- Exercise: Astronauts must exercise for at least two hours every day in order to stay fit and healthy. They do this by using special exercise machines like a stationary bicycle or a rowing machine.
- Energy: Electricity is needed to provide power for different equipment. This comes from solar cells which are used to charge batteries.
- Rubbish: Most rubbish is brought back to the Earth but sometimes large pieces of old equipment are thrown into space where they orbit the Earth and then fall into the atmosphere and burn up.
- Spacesuits: Astronauts must wear a spacesuit when they leave the space shuttle or space station to work outside. The suit provides the right pressure for the human body, provides oxygen, removes carbon dioxide and controls the temperature and humidity. They can also wear a disposable nappy in case they need to go to the toilet.

For you to try

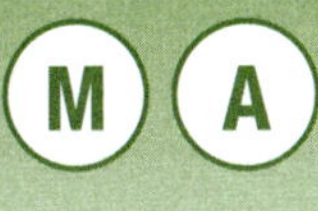

1. Write a story to describe how people have explored space in the last fifty years. (L) (A)
2. Draw a timeline to show some of the important things that have happened in the exploration of space. (M) (A)
3. Draw a diagram or picture of the information you would include in a metal plate like the one that was fixed to Pioneer, and may be found by intelligent living things. Write some notes around the outside of the drawing to explain why you included the different parts of your drawing.
4. Imagine that you are living in a space station where there is no gravity. What would life be like? How do you eat and drink, take exercise, have a bath and go to the toilet? Make a timetable to show the amount of time that you spend on each activity in one day. (L)
5. Make a model of a satellite, the space shuttle or a space station.

Looking into space

The Sun, Moon, planets and stars that people observe moving in the sky are known as **heavenly bodies**. Because they move in a regular way, people have learned how to use the movement of heavenly bodies. People in Papua New Guinea know how to use the sky as a clock, calendar and compass to navigate. The scientific study of the movement of the Sun, Moon, planets and stars is called astronomy. Astronomers use telescopes to study heavenly bodies that are far away. The first telescopes used glass lenses to magnify the light from the stars. Light telescopes can also be made with mirrors instead of lenses.

A simple telescope

The most powerful telescopes use radio waves instead of light and look like big satellite dishes. Radio telescopes are used to collect information from satellites and space probes. There is a famous radio telescope at Parkes in New South Wales, Australia. This is the dish that sent very clear pictures of the Apollo 11 Moon landing to the rest of the world in 1969.

The radio telescope at Parkes, New South Wales, Australia

For you to try

1 **Investigation: Making a simple telescope**

a Collect two cardboard or plastic tubes with different diameters and two different convex lenses or magnifying glasses (convex lenses are thicker in the centre than at the edge). You will also need plasticine or sticky tape to fix the lenses in the tube and polystyrene foam blocks to hold the smaller tube inside the bigger tube.

b Use the smallest, fattest lens as a magnifying glass to look at a page in a book. Move the lens up and down until the page is clearly in focus and measure this distance. This is called the **focal length** of the lens.

c Repeat step (b) with the bigger, thinner lens and again measure the focal length.

d Add the two focal lengths together. This will be the overall length of the telescope.

e Fix the smaller fatter lens into one end of the smaller tube using plasticine or sticky tape.

f Fix the bigger lens into one end of the bigger tube.

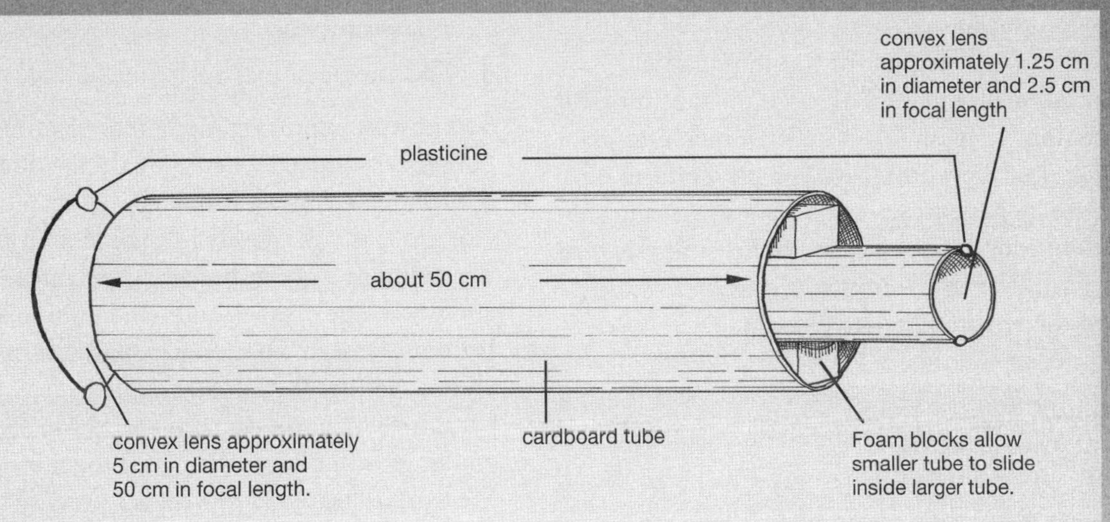

A simple telescope

- **g** Slide the smaller tube into the bigger tube and support it with blocks of polystyrene.
- **h** Make sure that the lenses line up with each other. The centre of each lens should be in the centre of the tube.
- **i** You can now use your telescope to look at distant objects. Slide the smaller tube in and out to make the image clear. The image will be clear when the distance between the two lenses is the same as in step (d).
- **j** The image will appear to be upside down. Most telescopes that you buy use more than two lenses so that the image then appears the right way round.

2 Investigation: Observing the Sun

- **a** You can observe the Sun by holding your telescope so the image falls on a piece of white paper or cardboard.
- **b** Never look at the Sun with your telescope as it is bright enough to damage your eyes.

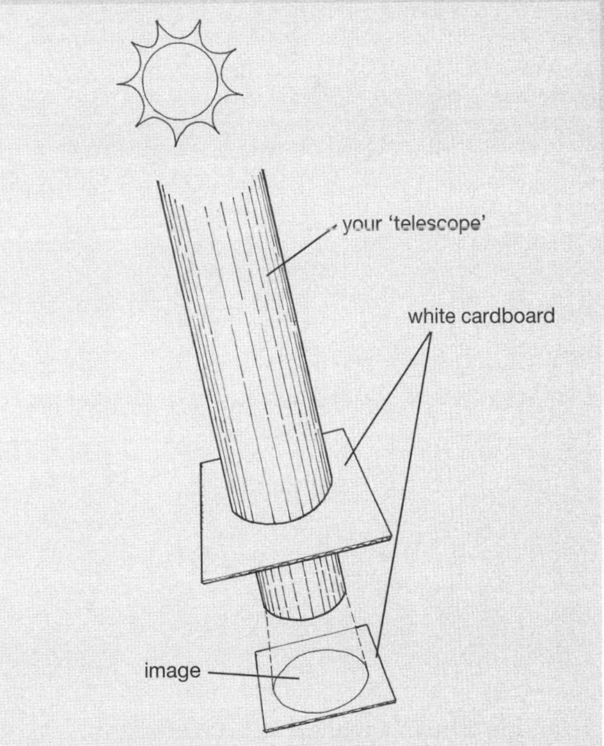

How to observe the Sun safely

Communication

Satellites are important in communication. Geostationary satellites are used to transmit radio, television, telephone and Internet messages to any part of the world. Messages in the form of electromagnetic waves can be transmitted to the satellite which can then send them to any part of the Earth's surface. The waves travel very quickly from the Earth to the satellite and back.

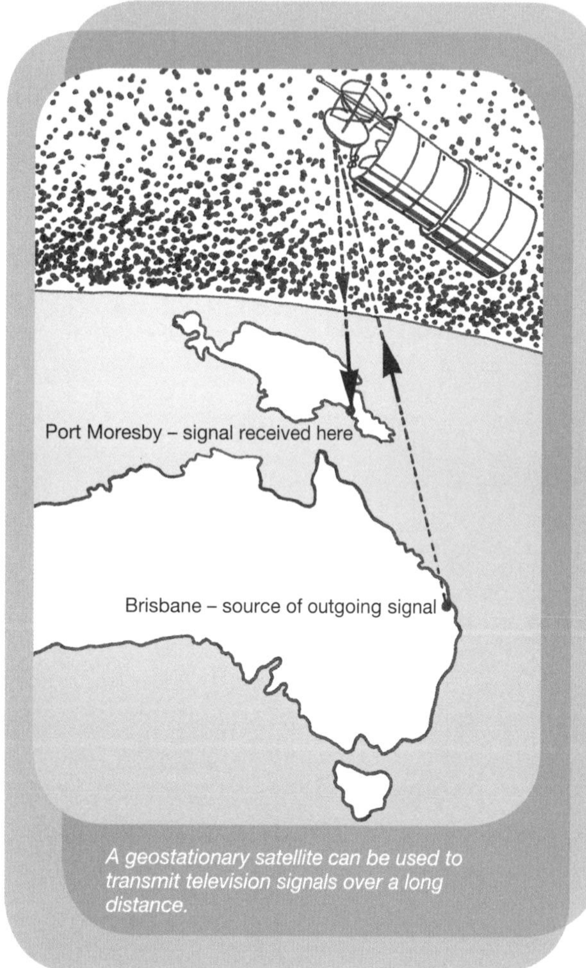

A geostationary satellite can be used to transmit television signals over a long distance.

There are three steps involved in all methods of communication:

1 The message must be sent by someone or something.

2 The message must travel to another place or be transmitted.

3 The message must be received by the receiver.

Sound is an important method of communication. Sound is produced when something vibrates. We can hear musical instruments because of the vibrations they make. For example, we can hear the sound of a guitar because the strings vibrate and we can hear the sound of a kundu drum because the skin vibrates. The particles of air also vibrate, so that the air is 'squashed' and 'stretched' continually. Sound can also pass through solids and liquids but it cannot pass if there is no air. We cannot hear sounds in space because there is no air.

Talking and hearing

We can make sounds by talking, singing, shouting and laughing. Inside your throat is a tube called the windpipe that carries air from your mouth to your lungs. Inside the windpipe are two flaps of skin called vocal cords. These are contained in a sort of box called the larynx or voice box. This is just behind the lump in your throat called the Adam's apple. When you want to make a sound your brain sends a message which pulls your vocal cords partly across the windpipe. When you breathe out the moving air pushes against the cords in your windpipe and makes them vibrate. You can make high or low sounds, loud or soft sounds by controlling your vocal cords. Your lips, teeth and tongue also help to make the sounds of the alphabet and of different languages.

When something like a guitar string or the skin of a drum vibrates it also makes the air vibrate which causes sound waves. The sound waves enter your ears and make your eardrums vibrate. Three small bones pass the vibrations on to the cochlea which contains liquid and has special nerve endings. The nerves then send messages to the brain which then sorts out the messages as different sounds.

Telephone, radio and television

Messages can be made to travel over long distances in several ways:

- Telephone systems transmit the message using electric currents in wires. The small dishes that can be seen outside Post Offices and on hills in the bush are used to transmit and receive telephone calls using the microwave system.
- Radio transmits the message using electromagnetic waves through an aerial at the radio station and is picked up by the aerial of the radio receiver.
- Television pictures can also be transmitted by electromagnetic waves. The signal is

tongue
vibrating vocal cords inside voico box
windpipe
moving air from lungs
larynx
oesophagus (food pipe)

Parts of the body used to change sounds into speech

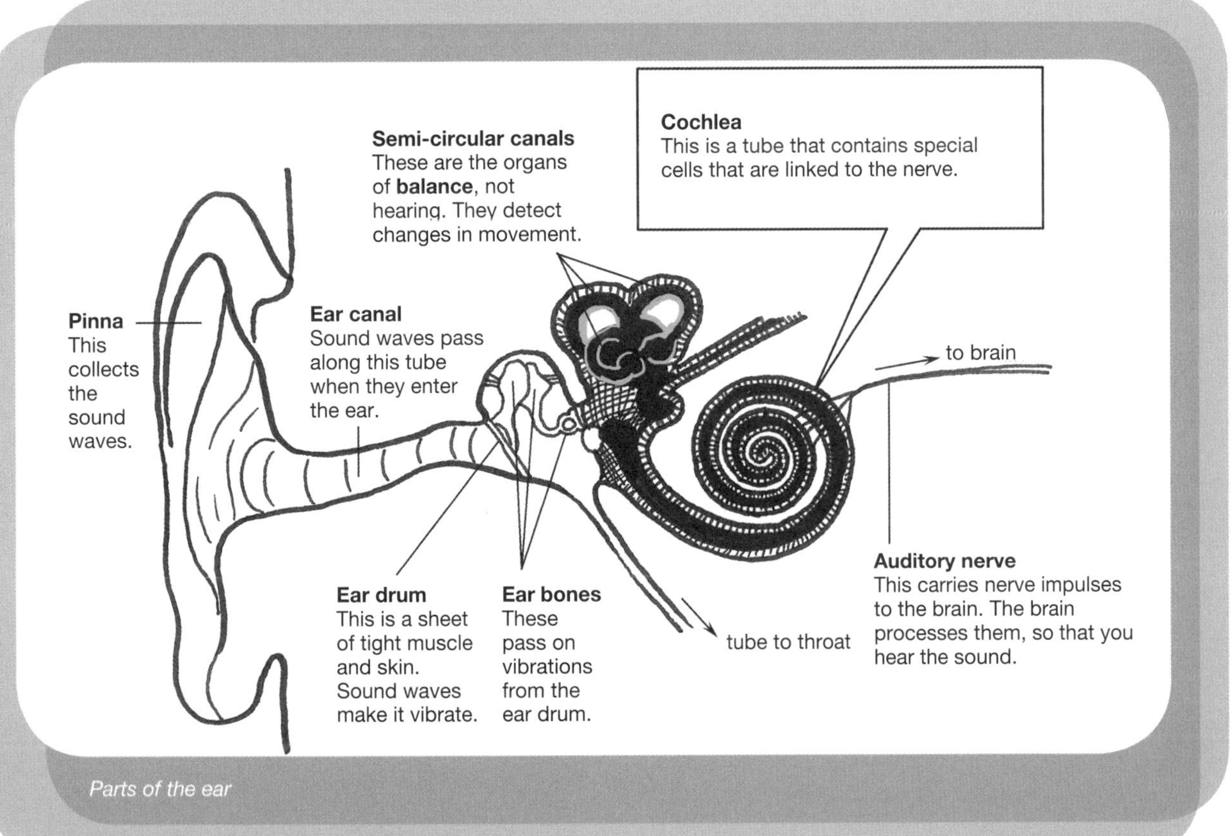

Parts of the ear

transmitted and received in the same way as for radio, but in this case the message includes information about the picture as well as the sound.

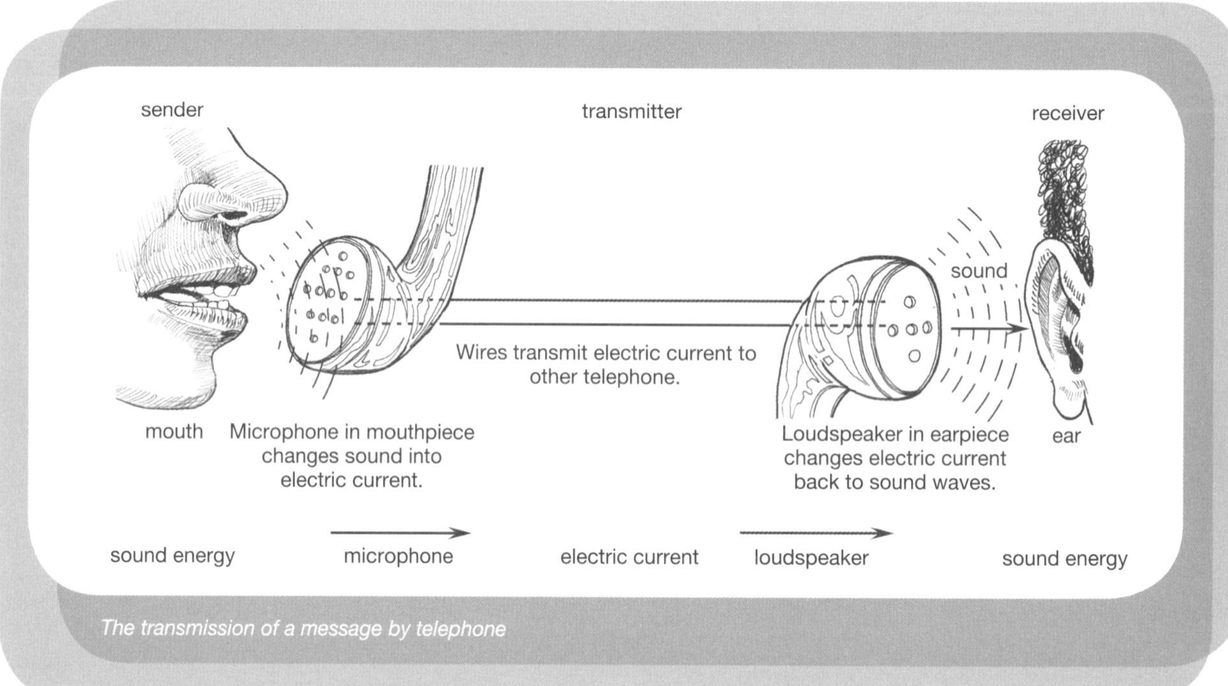

The transmission of a message by telephone

The dishes at Boroko Post Office

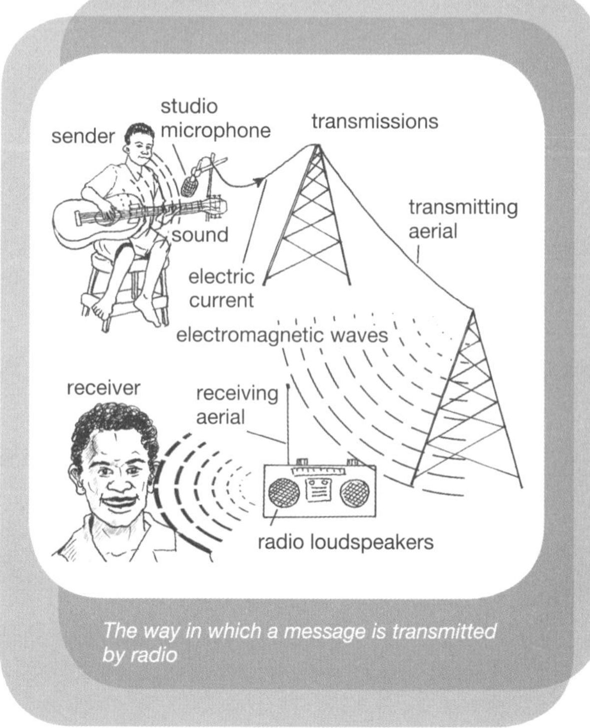

The way in which a message is transmitted by radio

For you to try

1 Investigation: Making a string telephone 

a Collect two clean tin cans and a length of string. You will also need a hammer and a nail.

b Make a small hole in the centre of the bottom of each tin can.

c Thread one end of the string through each hole and tie a knot on the inside of the can so that it cannot come out.

d Two people must hold the tins apart so that the string is always tight. One person talks into the tin can while the other person holds the other tin can to their ear.

e Does pulling the string tighter make the sound different?

f Try using different lengths of string. What difference does this make to the sound? How long is the string before you cannot hear your friend talking?

g Can you make the telephone work if your friend is round a corner?

h Write up a scientific report to explain what you did and what you found out.

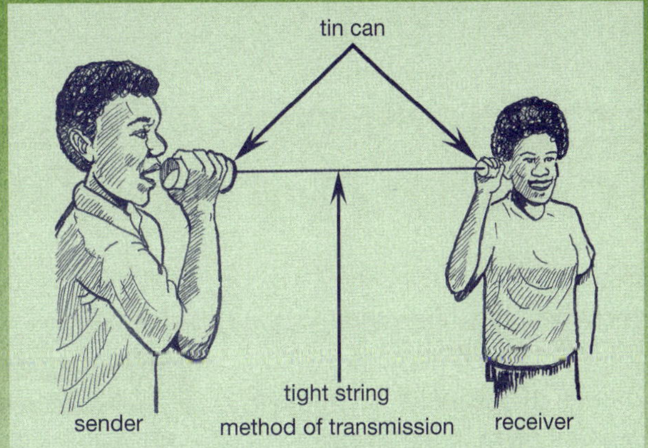

A string telephone

Projects

1 Collect a large sheet of paper and draw a timeline to show some important events in the exploration of space. Draw pictures of the different space craft that were used to help illustrate your timeline. Make your timeline as big as you can so that it can be displayed on the wall. 

2 Collect stories about satellites and space exploration from newspapers and magazines. Make a scrapbook or poster to display these stories. Have a display of scrapbooks and posters. 

3 Make a list of rules that astronauts must follow when they are living in space. Try to keep the rules as short as possible so they are easy to understand and follow. If you can make people smile when they read the rule it may help them to remember the rule. Make a poster to display your rules. You can also include drawings to help explain the rule. (L)

Summary questions

1 Which of the following best describes the order of the stages in which humans learned to use rocks and metals in some parts of the world?

- **A** stone age, bronze age, iron age
- **B** stone age, iron age, bronze age
- **C** iron age, bronze age, stone age
- **D** bronze age, stone age, iron age

2 Which of the following statements about minerals are true and which are false?

Statement about minerals	T or F?
Minerals are the chemical elements and chemical compounds that make up rocks.	
There are many different minerals and all are easy to find.	
Most rocks are made of minerals but some are not.	
The way a rock looks and its properties depend on the minerals that make up the rock.	
Most rocks are a mixture of a few minerals.	
Landowners in PNG also own all the minerals below the ground.	

3 The diagrams show the crystal size of igneous rock that cooled down at three different speeds.

Which of the following best describes the order in which the magma cooled, from the fastest to the slowest?

A I, II, III

B I, III, II

C III, II, I

D II, III, I

Questions 4 and 5 refer to the following information:

	Original rock		*Metamorphic rock*
I	limestone	→	marble
II	shale (mudstone)	→	slate

4 Which of the following best describes the reason for the change in I?

A heat

B light

C pressure

D metamorphosis

5 Which of the following best describes the reason for the change in II?

A heat

B light

C pressure

D metamorphosis

6 In the answer column in the table below match the type of rock (A to D) with the best description and examples of its use, I to IV.

	Description	Examples of use	Rock
I	very hard, sparkles due to minerals	building, making roads	
II	light colour, not hard	building, used in cement and concrete	
III	light colour, hard, smooth	building, statues	
IV	hard but splits into flat sheets	roofing tiles, pool table	

A limestone

B marble

C granite

D slate

7 Which of the following statements are true and which are false?

Statement	T or F?
Igneous rock comes from magma that has cooled and solidified.	
Different igneous rocks found in different parts of the world contain different minerals.	
The type of minerals they contain and the size of the crystals can be used to classify igneous rocks.	
Igneous rocks with small crystals have cooled slowly.	
Igneous rocks with large crystals have cooled quickly.	

8 The table below gives information about the formation of three kinds of sedimentary rocks found in different places.

Part of river	headwater	middle	coastal
Stage	young	mature	old
Sediments	gravel stones	sand	silt
Illustration			

Which of the following best matches the kind of sedimentary rock with the sediments described in the table above?

A	conglomerate	sandstone	shale
B	sandstone	conglomerate	shale
C	conglomerate	shale	sandstone
D	shale	sandstone	conglomerate

9 Which of the following would be good exercise for an astronaut living in a space station?

I riding a stationary exercise bicycle

II using a rowing machine

III lifting heavy weights

- **A** I and II only
- **B** I and III only
- **C** II and III only
- **D** I, II and III

10 For astronauts living in space which of the following activities are likely to be done differently than when they are on the Earth **because there is no gravity**? Answer true or false.

Activity affected by gravity?	T or F?
Using solar cells to obtain electricity	
Moving about and taking exercise	
Eating and drinking	
Having a bath	
Going to the toilet	
Bringing all their rubbish back to the Earth	

11 Which of the following best describes the order of events in which humans have explored space?

- **A** living in space stations, satellites and probes, people orbiting the Earth
- **B** people orbiting the Earth, satellites and probes, living in space stations
- **C** people landing on the Moon, living in space stations, people orbiting the Earth
- **D** people orbiting the Earth, people landing on the Moon, living in space stations

12 The passage below is a summary of the main ideas of this chapter. Copy and complete the passage in your book. Using the word in the list, find the words that are missing. You can use each word only once.

classified, communication, cooling, crystals, eroded, formed, heat, metamorphic, minerals, Moon, pressure, rock, satellites, size, space, ores

All rocks are made of ______________________. The way a ______________________ looks and its properties depend on the minerals that make the rock. Rock can be ______________________ in different ways—igneous, sedimentary and metamorphic. Igneous rocks are formed by the __________ of molten rock called magma. Slow cooling produces large ______________________, while fast cooling produces small crystals. Igneous rocks can be classified according to crystal ______________________, minerals and colour. Sedimentary rocks are formed from broken down rock that is ______________________ and carried by rivers to the sea where it collects and becomes cemented together again.

______________________ rocks are rocks that have been changed. Metamorphic rocks can be formed by ______________________ and pressure. The type of metamorphic rock produced depends on the amount of heat, the amount of ______________________ and the type of original rock.

Rocks can be ______________________ according to the minerals they contain, the size of the crystals and the way that we use them. Metals can be obtained from minerals and the extraction of metals from rocks known as ______________________ is important in Papua New Guinea.

We use ______________________ orbiting the Earth in space to transmit radio, television and telephone messages to any part of the world. Satellites are important in ______________________. We can find out more about ______________________ by using different kinds of telescopes. People have travelled through space to the ______________________ and there is an international space station orbiting the Earth which a few people from different countries visit for short periods of time.

Glossary

acids	a family of substances that contain the element hydrogen and have a sour taste; strong acids can be very corrosive
algae	simple green plants that usually live in water or damp places e.g. seaweed
alkali	a base that is soluble in water
alloy	a metal made by mixing two or more different metals
alluvial gold	small pieces of gold carried by rivers where it settles on the bottom
amphibian	an animal that is able to live on land and in water by breathing through the skin and lungs e.g. frog
andesite	a kind of igneous rock
anther	the structure from which pollen is released
armature	a rotating electromagnet in an electric motor
asexual reproduction	a type of reproduction in which a new individual is produced from one parent
atmosphere	the layer of gases that surrounds the Earth
atmospheric pressure	the weight of air pushing down on the Earth
axis	an imaginary line through the centre of the Earth; the line on the left hand side or bottom of a graph
basalt	a kind of igneous rock
bases	a family of substances that react with acids; strong bases can be very corrosive
binary fission	the process one cell divides into two cells and the contents of the two cells are shared
biodegradable	materials that can rot or be broken down by decay due to small living things like bacteria and fungi
block and tackle	a block of pulleys together with the rope or chain
brass	an alloy made from copper and zinc
brittle	hard but breaks easily
bronze	an alloy made from copper and tin
budding	asexual reproduction by forming a bud that slowly develops into a new individual
burr	hooks on the coat of a seed that help them to stick to the fur of animals
carpel	the structure that contains the ovary in plants
casting	pouring liquid metal or plastic into a mould to make an object with a particular shape
Celsius	a scale from 0 to 100 used to measure temperature
chemical properties	the characteristics of a material and how it reacts with other substances
circuit	a path around which electricity can flow that is usually made of wires
clearfelling	cutting down all the trees in the forest
cocoon	a special case that many insects spin at the end of the larval stage
communication	the process in which a message is sent, transmitted and received
commutator	part of an electric motor that reverses the direction of the current in the armature so that the armature continues to rotate in one direction
conductor	something that lets heat or electricity pass through easily
conduction	the way in which heat moves in a solid
conserve	to use materials carefully so that they will last as long as possible
contraction	the decrease in size that occurs when some substances are cooled
control	the part of an experiment in which no changes are made so that the result can be compared with the test; *see also* test
convection	the movement of heat that occurs when hot air expands and floats upwards and cooler air moves in to take its place
corrosion	a process that changes and spoils metals e.g. rusting
corrosive	something that can cause serious burns to the skin and also make holes in clothes e.g. strong acids and strong bases
cross pollination	pollen that is carried from stamens to carpels on another plant of the same type
data	information that has been collected in an investigation
degree	a unit of temperature
density	the heaviness of an object for its size; the amount of material in a given space
dispersal	the scattering or spreading of seeds
dormant	a seed that is resting
drift net	a long net that is allowed to float behind a boat in order to catch fish
durable	does not wear out easily
edible	something that can be eaten
efficiency	the energy that we get out divided by the energy that we put in multiplied by 100%
egg	a female sex cell
electrodes	metal plates that dip into a liquid and are used to pass an electric current through the liquid
electromagnet	the magnetic effect that is created when a electric current flows in a coil
electromagnetic radiation	the way in which heat and light travel from the Sun
embryo	a zygote that has divided and grown

energy energy is the ability to do work; there are different types of energy

exoskeleton the outside covering of insects, prawns and crabs

expansion the increase in size that occurs when some substances are heated

external fertilisation outside the body of the female

extrusive rocks igneous rocks that are formed when magma reaches the surface of the Earth

fair sensible, reasonable, logical

fern a green plant that has no flowers but reproduces by producing spores on the underside of the fronds

fertilisation the joining of a male sex cell with a female sex cell

filament a wire with high resistance

fissure a deep crack

flammable burns easily

flexible may be bent easily

flowering plant a plant that has enclosed seeds formed within a fruit that develops from a flower

foetus the developing baby from the third month of development until birth in humans

force a push or a pull that usually makes something move or change direction

fossil the remains of plants or animals that lived thousands of years ago

frogspawn frog eggs that have been laid in water

fruit a complete ovary after fertilisation

fulcrum a fixed point or pivot that is used to make a lever work

fungi plants that have no chlorophyll but live on the dead remains of other plants and animals

gas matter that has no fixed volume and no fixed shape; *see also* diffusion

geostationary satellite a satellite that appears to be stationary above one place on the Earth's equator

granite a kind of igneous rock

gravity the force that pulls objects down

hard difficult to scratch and not easy to wear away

heavenly bodies the Sun, Moon, planets and stars

horizontal parallel to the bottom of the page or horizon

hornfels a common type of metamorphic rock

hydro-electricity electricity made by generators driven by running water

igneous rock rock formed from magma that has cooled and solidified at the Earth's surface (volcanic rock) or deep within the earth's surface (plutonic rock)

imago the adult stage of complete metamorphosis in the life cycle of an insect

inedible something that should not be eaten

inclined plane a slope, such as a ramp, that allows heavy objects to be lifted or raised by smaller forces

indicator a dye that will change colour when mixed with an acid or a base

insulator something that stops heat or electricity passing through

internal fertilisation inside e.g. inside the body of the female

intrusive rocks igneous rocks that are formed inside the Earth

joule (J) a unit of work

kilojoule (kJ) a unit of energy, equal to 1000 joules; the symbol for kilojoule is kJ

kinetic energy moving energy

larva a young insect

life cycle the different stages a living goes through as it grows and develops from an egg to a mature adult

liquid matter that has a fixed volume but no fixed shape

litmus a common indicator

magma molten rock inside the earth

marble a metamorphic rock that is made from the sedimentary rock limestone

mass the quantity of substance or amount of material in something

matter any material or substance that has mass and takes up space

mechanical advantage a force ratio which we calculate by dividing the load by the effort

megajoule (mJ) a million joules

metal solids that are usually hard and can be flattened into sheets and stretched into wires

metamorphic rock rock that has been changed because of high temperature and pressure; it is harder and looks different

metamorphosis a process of change e.g. the separate stages that an insect goes through in its life cycle; during incomplete metamorphosis insects go through three stages: egg, larva or nymph and adult; during complete metamorphosis there are four separate stages: egg, larva, pupa and adult

mineral ores rocks that contain metals

minerals the chemical elements and chemical compounds that make up rocks

molten rock or metal that has been heated to a very high temperature and has become a thick sticky liquid

moss a simple green land plant that lives in cool, damp places

mould a shape or hole which is left by the body of an animal in mud, clay or ash; when the hole is filled it becomes a cast which has the same shape as the original animal

natural materials materials that come from plants, animals and the Earth

neutralisation the type of reaction that occurs when an acid mixes with a base to form a neutral solution

non-biodegradable materials that do not rot or decay

non-renewable something that cannot easily be replaced

nymph a young insect that looks very much like the adult

obsidian a hard, shiny black igneous rock that can be used to make sharp tools

ore the rocks from which we obtain metals

ore body a large amount of ore

overfishing catching more fish than are born and grow to maturity

ovule a female sex cell or egg in plants

panning a way of obtaining gold by putting sand, gravel and water in a metal pan and moving it round in a circle

parallel circuit a circuit made from several loops in which the current divides and flows to each part separately

pasteurisation killing germs that cause food to become rotten by heating to 55°C

penis male reproductive organ in mammals that leaves the sperm inside the body of the female so that fertilisation can occur

pH scale a scale of numbers from 0 to 14 that is used to measure the strength of an acid or base
physical properties the characteristics of a material and how it behaves
pivot see fulcrum
placenta the structure that attaches the embryo to the uterus and which passes food and oxygen from the mother to the embryo and carries waste from the embryo to the mother
placer deposit the place where alluvial gold is found
pollen male sex cells in plants
pollination the movement of pollen from one flower to another flower
pollution materials that end up in rivers, the sea, on the land or in the air that can spoil the environment and harm living things
porous will allow water to pass through
processed materials materials that people have made or manufactured e.g. flour and cement
pulley a simple machine that consists of a rope, chain or belt stretched over the rim of a wheel
pupa third stage of complete metamorphosis in the life cycle of insects that does feed at all
quarantine special rules to make sure that plant and animal pests and diseases are not carried from country to country
radiation the way in which light and heat travel through air
raw materials naturally occurring materials that are used to make products
recycling using materials again instead of being thrown away; a way of conserving materials
regeneration asexual reproduction in which a new individual develops from pieces of the adult
renewable something that can be replaced
resource something that can be used to meet the needs of people
royalties money that is paid to landowners where timber or minerals are removed
satellite something that orbits or travels around the Sun
seed a fertilised cell that has grown by cell division in the ovary; each seed contains an embryo, a food store and a thick coat for protection
self-pollination pollen that is carried from stamens to carpels on the same plant
series circuit a simple circuit in which everything is connected in a line or single loop
sexual reproduction a type of reproduction in which a new individual is produced from two parents
sediment small particles of solid that settle at the bottom of a liquid
sedimentary rock rock that is formed from sediments that have been compressed
selective logging cutting down only certain types and sizes of trees
simple fission see binary fission
solar from the sun
solar system the Sun and the family of eight planets, moons and asteroids
solid matter that has a fixed volume and a fixed shape
space the empty place that surrounds or lies between the stars and planets
space probe a spacecraft that flies by or lands on the surface of a heavenly body in the solar system
space shuttle a space craft that takes off like a rocket, orbits the Earth and lands like an aeroplane
space station a satellite that is big enough for people to live in
spawning laying eggs in water e.g. fish and some frogs
sperm male sex cells in animals
spontaneous generation the idea or theory that living things just seemed to appear by themselves
spores single cells that are very light and usually produced in large numbers during asexual reproduction
stamen the structure that hold the grains of pollen that contain the male sex cells
state the condition of matter—whether it is a solid, liquid or gas
stem part of the shoot of a flowering plant that supports the leaves, buds, flowers and fruits
stock the animals or plants that are kept or left so that they will reproduce
synthetic materials processed materials that are made from chemicals or artificial substances
tadpole the embryo of a frog that develops from a fertilised egg
tailings the waste from ore that mining companies produce
telescope an instrument that is used to look at heavenly bodies
temperature a measure of how hot something is
terminals the parts of a battery to which the wires are connected
test the part of an experiment in which changes are made so that the result can be compared with the control; *see also* control
testes the place where sperm cells are produced in vertebrates
thermometer the instrument used to measure temperature
thermostat a switch that uses temperature to turn on and off
translucent lets light through but scatters it; opaque
transparent lets light through or 'see-through'
trawling catching fish or prawns by dragging a large net behind a boat
treated timber wood that has been soaked or sprayed with a chemical solution to help to stop it rotting or being eaten by insects
umbilical cord the tube that carries joins the embryo to the placenta
uterus the organ inside the body of a female placental mammal in which the embryo develops; also called the womb
vacuum flask a bottle with a double wall that is used to keep food and drinks hot or cold
vagina female reproductive organ in mammals that the penis enters and through which the baby passes when it is born
vegetative reproduction asexual reproduction in plants that occurs when a new individual is produced from any part except the flower
vertical at right angles to the bottom of the page or the horizon
waterproof will not allow water to pass through
weight the downward pull of the Earth's gravity on the mass of an object
zygote the single cell that is formed as a result of fertilisation